"A wonderful compendium and source for inspiration in the field of reducing and managing disaster risk. An immense contribution to an essential requirement for sustainable development in the world."

Salvano Briceno, *former Director of UNISDR (now UNDRR), 2001–2011.*

"This book reinforces inevitable principles from studying disaster risks: disasters are socially constructed processes; we must look at the global from the local and the local from the global; always privilege the identification of root causes of disasters. It brings together theoretically, methodologically, and conceptually the disaster risk realities experienced daily around the world."

Virginia García-Acosta, *Emeritus Researcher, Professor in History and Anthropology, CIESAS (Center for Research and Advanced Studies in Social Anthropology), Mexico.*

"This volume represents a good overall synthesis of existing literature, but also expands beyond traditional topics to include analyses of the COVID-19 pandemic and potential hazards beyond the Earth. This volume continues the extraordinary work of Ben Wisner and his colleagues, providing insights into existing and evolving hazards and suggesting pathways for action."

Michèle Companion, *Professor of Sociology, University of Colorado–Colorado Springs, and President of the Research Committee on Sociology of Disasters (RC-39) at the International Sociological Association (ISA).*

Understanding and Addressing Disaster Risk

In *Understanding and Addressing Disaster Risk*, the authors explain how people modify the environment and exert power over each other in ways that make nature potentially harmful and put people in harm's way. Opportunities and challenges faced by those engaging with disaster risk are explored.

Across 11 chapters, the authors show that disasters are not natural, are not events, and do not happen quickly. Instead, they are the result of chronic societal processes emerging from the creation and perpetuation of vulnerabilities and limitations on people's abilities to respond to hazards. The book also explores the environmental component of disaster risk through the lens of different natural elements and phenomena, including biological-ecological and water-weather-climate processes as well as geological and outer space dynamics. The authors explain the mutual influence of the different components of disasters in creating disaster risk across diverse regions of the world. They critique attempts to reduce disaster risk through top-down, siloed assumptions, attitudes, and values. The value of people's knowledge of hazards – often ignored or dismissed by authorities – is a central theme. This book is original because of how it re-interprets and advances understanding of the disaster process through the study of such societal processes of vulnerability, risk creation, and power imbalances. It is also unique in diving further into "root causes" of disaster in order to place them within local histories and colonial legacies as well as contemporary, typically misdirected, agendas while upending previous "solutions" which have been shown to do more harm than good.

Understanding and Addressing Disaster Risk is useful for and useable by decision-makers, policy makers, researchers, and students to shatter the vicious cycle of repeating known mistakes which compound detrimental outcomes.

Ben Wisner is an activist scholar who finds himself tempted by nostalgia for the 70s, 80s, and 90s when he worked to understand and address disaster risk with civil society and local government in a number of countries in eastern and southern Africa, Asia, Latin America, and the Caribbean.

Irasema Alcántara-Ayala is a professor and former director of the Institute of Geography at the National Autonomous University of Mexico.

JC Gaillard is Ahorangi/Professor of Geography at Waipapa Taumata Rau/The University of Auckland.

Ilan Kelman is Professor of Disasters and Health at University College London, England, and a Professor II at the University of Agder, Kristiansand, Norway.

Victor Marchezini is a sociologist at the Brazilian Early Warning Center (Cemaden).

Understanding and Addressing Disaster Risk

Who Speaks? Who Suffers?

Ben Wisner, Irasema Alcántara-Ayala,
JC Gaillard, Ilan Kelman, and Victor Marchezini

Routledge
Taylor & Francis Group
LONDON AND NEW YORK

Cover image: Photograph by JC Gaillard of Fogo, Cape Verde (18 April 2009)

First published 2025
by Routledge
4 Park Square, Milton Park, Abingdon, Oxon OX14 4RN

and by Routledge
605 Third Avenue, New York, NY 10158

Routledge is an imprint of the Taylor & Francis Group, an informa business

With thanks to University College London's Open Access team for covering the fee required to make this book freely downloadable.

British Library Cataloguing-in-Publication Data
A catalogue record for this book is available from the British Library

ISBN: 978-1-032-27444-7 (hbk)
ISBN: 978-1-032-27445-4 (pbk)
ISBN: 978-1-003-29281-4 (ebk)

DOI: 10.4324/9781003292814

Typeset in Univers LT Std
by Apex CoVantage, LLC

With thanks to University College London's Open Access team for covering the fee required to make this book freely downloadable.

Contents

Illustrations

Figures

Tables

About the Authors

Ben Wisner is an activist scholar who finds himself tempted by nostalgia for the 70s, 80s, and 90s when he worked to understand and address disaster risk with civil society and local government in a number of countries in eastern and southern Africa, Asia, Latin America, and the Caribbean. In his 8th decade, he watches the autumn leaves turn color, contemplates the cycle of life, death, and rebirth, and wonders what impact his university teaching and publications have had. The warm southwest wind whispers the names of his many co-authors, co-editors, and research partners. He is grateful for them all. A detailed autobiography and samples of his published papers, together with critical commentaries and his replies, will be available in 2025 in a Routledge volume called *Reflecting on Ben Wisner's contributions to scholarship and scholars: The Benschrift* edited by JC Gaillard and Ilan Kelman.

Irasema Alcántara-Ayala is a Professor and former director at the Institute of Geography and at the National Autonomous University of Mexico (UNAM). Born in Mexico City, she earned her bachelor's degree in geography from UNAM and her doctorate from King's College, London, and later completed a postdoctoral fellowship at MIT. Irasema is dedicated to studying geomorphology and disaster risk, focusing on connecting scientific knowledge to practical solutions in disaster management. She contributes to various scientific committees, including the Scientific and Technical Advisory Group of the Regional Office of the United Nations Disaster Risk Reduction for the Americas and the Caribbean, striving to support disaster risk reduction and management efforts.

JC Gaillard is Ahorangi/Professor of Geography at Waipapa Taumata Rau/The University of Auckland. His work focuses on power and inclusion in disaster and disaster studies. It includes developing participatory tools for engaging minority groups in disaster risk reduction with an emphasis on cultural and gender minorities, people in detention, and children. More details from https://jcgaillard.wordpress.com.

Ilan Kelman https://www.ilankelman.org/ and Instagram/Threads/X @ILANKELMAN is Professor of Disasters and Health at University College London, England, and Professor II at the University of Agder, Kristiansand, Norway. His overall research interest is linking disasters and health, integrating climate change into both. Three main areas are as follows: (i) disaster diplomacy and health diplomacy (https://www.disasterdiplomacy.org/), (ii) island sustainability focusing on safe and healthy living and livelihoods (https://www.islandvulnerability.org/), and (iii) risk education for health and disasters (https://www.riskred.org/).

Victor Marchezini is a sociologist at the Brazilian Early Warning Center (Cemaden). He is a professor at the Doctorate Program on Earth System Science, National Institute for Space Research (PGCST/INPE), and at the Graduate Program on Disaster (a joint initiative between ICT/Unesp and Cemaden). He is coordinating the project COPE (Fapesp grant number #2022/02891–9).

Foreword

Understanding and Addressing Disaster Risk is a welcome and important addition to writing about disasters. The book reviews definitions and implications of key concepts: hazard, risk, vulnerability, coping, and abilities. The authors strengthen the concept that disasters are not natural, a notion promoted in contemporary science as early as the 1970s by social scientists and followed through by progressive writers, including the authors of this book ever since. However, the notion of "*natural disaster*" has persisted in institutional writings, including through different UN agencies even today. This led Mami Mizutori, the penultimate Special Representative of United Nations Secretary General for Disaster Risk Reduction and head of the United Nations Office for Disaster Risk Reduction (UNDRR) to launch a "No Natural Disasters Campaign" so as not to promote the idea that "*disasters are inevitable, and that human agency can do little to prevent or mitigate their impacts*" (UNDRR, 2020).

This is a book based on deep-rooted philosophical, ethical, moral and human rights principles which, in its arguments, seek to reconnect us and the human world with our mutually-dependent counterpart, the non-human biotic and inorganic world. It recognizes the cultural, and thus qualitative and interpretative, differences in the significance and meaning of disaster risk and disaster. The authors seek to move academia and research endeavors toward a paradigm for promoting change and action. Not content merely to understand risk and vulnerability, the authors promote intervention based on culturally diverse, local, and community expressions of life and life needs. The book reconnects disaster risk reduction to the search for happiness and security in a wider sense. In sum, it is a treatise that attempts to see disaster in the light of everyday life, its challenges, and needs. It places intervention or action within the context of socially participatory schemes and links DRR to the efforts of social movements to bring about greater equality, balance with nature, and social justice. By challenging the way that normal "development" often creates risk, the book is also a profound challenge to status quo policy and practice.

The timeliness and relevance of this book is obvious when one considers the lack of progress achieved in reducing disaster risk over the past 50 years. First of all, disaster risk continues to grow, loss and damage are constantly increasing, and systemic effects are amplifying. The evidence is indisputable as shown by statements over the last ten years since the enactment of the Sendai Framework for Disaster Risk Reduction, from academia, the United Nations, World Bank, the European Union, insurance companies, NGOs, and government agencies. Meanwhile, attention to disaster risk and investment in DRR is stagnant on global, national, and local levels. The direct and indirect impacts of disasters are substantial in terms of eroded livelihood opportunities, reduced development potential, wasted resources and investment, and overall lowered societal sustainability and ability to achieve the Sustainable Development Goals (SDGs). This situation is even more poignant when one considers that it is in the poorer countries and amongst more disadvantaged and excluded populations that disaster risk is most acute.

Secondly, increased loss and damage occur under increasingly complex hazard and risk conditions. Risk drivers are increasing and diversifying with climate change and economic and political globalization. Local communities and even whole countries are more and more affected by processes out of their control, reach, and jurisdiction. Overall, at the local level, awareness of basic climate science and its impact on everyday existence is low. Access to technologies to integrate climate factors to the planning process and access to information is limited.

Third, DRR is still conceived, in most places, as a discrete sector among other sectors and not an integral, cross-cutting component of development thought and practice, part of its DNA. There has not been a fundamental break with the past and an institutional reformulation under new governance principles since disasters emerged as a modern public concern more than 70 years ago. This cognitive and institutional inheritance concentrates attention on response and preparedness, where a collaborative, sectoral perspective is thought to be adequate to incorporating a range of actors located in different institutional and organizational settings. DRR becomes an "add-on" to development practice and not as an essential and defining component when it is limited to a sectoral format. DRR receives less attention when competing for scarce investment resources.

Obstacles to overcoming DRR's isolation and neglect include a lack of political commitment to address the root causes and immediate drivers of disaster risk based on recognition that skewed development and economic growth are the basis of disaster risk creation. DRM has thus stayed in the domain of the relief structures, a process encouraged and demanded by each crisis and disaster. Under these conditions, finance and investment fall far short of response needs and agency requests. Options for non-governmental-led DRR, based on civil society, are limited due to problems of lack of access to decision-making and to financial and human resources for coping, resilience, and overall sustainability.

Fourth, the need for DRR to be thought through and practiced as a fundamental aspect of sustainable development requires understanding that disaster risk is not a break with quotidian life and risk, an exogenously generated condition, but rather is the result of internal, endogenous processes that lead eventually to disaster risk and disaster because of their effect on daily life and access to resources. The "normal" disaster risk context with low-impact, frequent disaster occurrences eventually leads to more severe disaster scenarios. Disaster risks are expressions of the limits of our relationship with that "other" nature on which we depend. Disaster is a vivid representation of the breakdown in harmonious ecological and human relations. Domination of nature as opposed to harmony and mutual respect prevails and has done so increasingly over the last centuries and millennia. This toxic relationship with nature is not considered in planning strategies.

Fifth, currently no widespread action to integrate DRR and climate change adaptation (CCA) exists. One result is a tendency to return to hazard-based interpretations of loss and damage, of disaster and calamity, with the relegation to the background of vulnerability and exposure and, hence, neglect of close consideration of root causes and more immediate risk drivers. Further, there is only limited linkage between DRM/DRR and practices that would build climate resilience. This is accompanied the sectorization of CCA, isolating it from other hazards. The dominance of discourses around extreme events, climate risk, climate mobility, climate crises, and other hazard-dominated terminologies has set us back and shows how the narrative molds the action framework and the consideration of what is important and a priority. In addition, governments do not have sufficient political will to tackle climate change. "Climate crisis" should be expressed as "societal crisis" under the effects of climate hazards; climate as such is never in "crisis", although it may be, and has been, changing, in flux, or unstable. Similarly, climate mobility, as a notion, fails to call our attention to the many social, economic, political, and cultural influences that drive mobility.

And sixth, a clear understanding often does not exist that disaster risk is local in expression, although global, national, and subnational development processes directly influence localities. Thus, communities are limited in their ability to act to safeguard their lives and livelihoods due to over-centralization of decision-making and finance, as well as the lack of consideration and incorporation of local actors. Community-based DRR approaches are not incorporated or considered in planning and implementation processes. Or, where considered, the participation and involvement of people is many times more of a lip service.

Understanding and Addressing Disaster Risk addresses these deficiencies and disappointing failures. This not just another in a long parade of books that skim the surface and look away from misuse of power. It is uncompromising and humane. The authors are also thorough. There are chapters that engage with living and non-living nature, the geological, geomorphological, and biological realities on which our lives and livelihoods depend, and also the sources of the earthly hazards we confront. For all these reasons, this book's reading is essential.

Zenaida Delica-Willison and Allan Lavell
Center for Disaster Preparedness, Philippines (for ZDW)
Facultad Latinoamericana de Ciencias Sociales, Costa Rica (for AL)

Homage

From this book's second, third, fourth, and fifth authors.

In celebration of this book's first author's 80th birthday and his unwavering dedication to reducing disaster risk, *Understanding and Addressing Disaster Risk* is both a tribute and a testament to a remarkable journey of more than half a century. Ben's passion has driven groundbreaking research and advocacy across disaster risk, community health, applied nutrition and food security, child welfare, and water and sanitation. His pioneering use of participatory methodologies has advanced scientific understanding and tremendously impacted countless communities' lives, enabling them to take charge of their safety and well-being. Ben's unique integration of Buddhist philosophy and the profound role of emotions, like compassion, anger, regret, and shame, brings a profoundly human dimension to research and ethics, emphasizing their critical importance in meaningful and ethical communication of scientific results. His invaluable contributions have deeply influenced several generations of disaster risk researchers and practitioners. Ben's steadfast commitment, boundless zest, and visionary insights have left an indelible mark on the field, inspiring all who have had the privilege of learning from him to strive for a safer and more secure world. This book honors Ben's extraordinary contributions, offering both a heartfelt tribute to his life's work and a comprehensive, scientific exploration into the complex factors that shape disaster risk and people's well-being.

Preface: A Long and Hazardous Road

This book has a story. It has five authors, and they also have stories. These stories combine to map the long and hazardous road this book has followed. "Long" because the book's origins reach back to the late 1980s, when a series of meetings in London, Oxford, and Norwich in the UK crystallized the idea of *At Risk* (finally appearing in 1994). "Hazardous" because a team of five authors must juggle many responsibilities, time tables, periods of ill health, and other misadventures. It is also only by perseverance and good luck that five personalities don't clash but work together as smoothly as we have.

Individual Stories

Ben grew up in Northern California, where, as an adolescent, he experienced moderate earthquakes and summer season wildfires in the hills adjacent to his home. Later, he lived in a Tanzanian village between 1966 and 1968, and he saw close at hand the hazards faced by his neighbors: house fires that could spread quickly in the densely packed settlement, flooding, crop pests, livestock disease, malaria, and many other diseases. Later, he used the Swahili he had learned to discuss with farmers how they coped with drought on land on the lower slopes of Mount Kenya. In particular, he talked with them about how they tried to protect their small children from the impacts of drought. Thus health, health care and the lack of it, diet, and nutrition were the foundations on which Ben's curiosity about hazards, risk, and especially vulnerability were built. They are also the origin of his passion to reduce human suffering by applying the knowledge gained.

Inspired by her mother's example, **Irasema** wanted to pursue a career as a teacher and found herself in a high school literature class when the Mexico City earthquake struck on 19 September 1985. Although her school remained unharmed, it was located near the disaster's devastation zone, prompting Irasema to volunteer in relief efforts. Later, during college, a visit to a landslide-affected town highlighted the lack of experts in Mexico in mass movement processes and disaster studies. This gap in knowledge and expertise, coupled with her fascination with the magic of landscapes, particularly volcanoes and mountains, and her geographical spirit, fueled her interest in the nexus between people and the Earth's surface, processes, and landforms to better understand disaster risk.

Her academic and applied work earlier led her to direct the Institute of Geography at the Autonomous National University of Mexico, where she is currently Professor of Natural Hazards and Risk.

JC's one home in France, Passy, is nestled within a hazardous landscape, prone to floods, debris flows, landslides, mass movements, rock falls, and avalanches. In 1751, the collapse of part of the mountain made the king in Turin think that there was a volcano erupting. In 1970, 72 people (including 56 children) died in debris flows. JC guesses that this hazardscape triggered his original interest in studying disaster. This interest was further stimulated when JC lived in his other home, the Indung Kapampangan, Philippines, experiencing several cyclones and magnitude 6+ earthquakes while relying on a limited government salary with no benefits, hence, experiencing some dimensions of vulnerability firsthand while still feeling privileged in many ways.

Ilan began in disaster research when he frequently saw headlines and news reports of disasters, with little explanation of why they happened and even less information on examples where disasters did not happen, explaining why they did not happen. He was further concerned that disasters seemed to set back people and places from improving their living conditions. He started wondering how disasters could be stopped, especially since so many people seemed to care about disasters only after they had happened, when it was too late. This wondering led to a master's degree, work in Caribbean development, and then a PhD, followed by a career trajectory in practical and policy-relevant academia.

Victor was part of a collective that initiated the sociology of disasters in Brazil, analyzing disaster impacts and creating participatory methods to enhance people's capacities, especially children, with whom he found it exciting to work. This collective also revealed the human rights violations in the "temporary" shelters and identified forms of abandonment as part of the biopolitics of disasters. He felt angry about this. When he started a new position as a disaster analyst at the Brazilian Early Warning System, he was intrigued by the recurrent question from his coworkers: "What is a sociologist doing here?"

It is perhaps significant that four of the five co-authors had early experience with children – those in school settings (Wisner, 2006) and, more widely, with children who suffer in disasters or act as agents in disaster reduction (e.g., through participatory three-dimensional disaster mapping, P3DM). Also of note, all five eventually applied their curiosity and knowledge gained in practical ways to reduce hazard impacts and people's vulnerability in a variety of institutional settings.

Networks of Stories

Besides the stories of each of its authors, the book itself was inspired and grew from roots nourished by two editions of *At Risk*. Ben was one of the co-authors of *At Risk* together with Ian Davis, Piers Blaikie, and Terry Cannon. Our present book has been influenced indirectly by other work published by Ian, Piers, and Terry alongside many venues for discussion, in-person and online around the world. Each of the five authors is part of overlapping groups, such as the Network for Social Studies on Disasters in Latin America (La Red https://www.desenredando.org) and the #NoNaturalDisasters #DisastersAreNotNatural campaign (https://www.radixonline.org/). Among our individual collaborators are hundreds of scholars, policy makers, and practitioners, notably the authors of this book's Foreword and everyone contributing to the Radix network (https://www.radixonline.org), founded by Ben and Maureen Fordham, followed by management, curation, and leadership by JC. Each of them, and so many others, have indirectly influenced the thinking of one or more of us.

More direct influence came from the writing of Ben's four current co-authors, including JC Gaillard's *The Invention of Disaster* (2021) and Ilan Kelman's *Disaster by Choice* (2020), as well as Victor's peer-reviewed papers on flooding in Brazil and on participatory early warning systems and his edited multilingual collection of work on the prevention of disaster in Brazil (Marchezini & Wisner, 2017; Marchezini et al., 2017). Not least is Irasema's prolific writing on geomorphological hazards, landslides in particular, forensic investigations of disaster, integrated community and environmental approaches to mountain hazards, and much more. The collections edited by three of the current co-authors contain the work of a wide range of authors whose work has been influential to the synthesis offered here (Wisner et al., 2012, 2015); whilst the definitive textbook written by another three of the present co-authors and others cite many sources and articulate a view of disasters that has influenced the whole of the present author team (Alcántara-Ayala et al., 2023).

What Has Changed

We would like to think that our present book is greater than the sum or product of its part origins. It is certainly not meant to be a third edition of *At Risk*, despite its family history. There are two reasons.

First, our understandings of the world's complexity have deepened and broadened since 1994, when *At Risk* (Blaikie et al., 1994) was first published. Indeed, its second edition in 2004 (Wisner et al., 2004) required an additional 187 pages to account for some of the changes that had taken place. As we write, yet another 20 years have elapsed. Consider the world in 2004. At the beginning of that year, Haiti still had an elected president, there was hope that peace and stability could be obtained in Iraq, South Africa celebrated ten years since the end of apartheid and many still had optimism about justice in the rainbow nation, and there were negotiations over the unification of Greek-and-Turkish-controlled Cyprus. Ten member states were added to the European Union that year. A humanitarian treaty ending the conflict in Sudan was agreed. International negotiations on environmental topics, including human-caused climate change, were ramping up following the optimism and energy of the 1992 Rio summit which founded three United Nations secretariats and presented Local Agenda 21. By 2024, that energy and optimism seem exhausted. The hoped-for reconciliations and peacemaking have not been forthcoming. Conflicts have proliferated, from Haiti descending into violence and ungovernability alongside South Africa's savior, the ANC, proving so corrupt and ineffective that they lost their majority in national elections. The gap between rich and poor has increased around the world.

Second, there has been a sea change in social theory that explains why our book is not meant to be a third edition of *At Risk*. A new generation of highly-talented and committed researchers well-versed in critical theories is pushing the boundaries of our field of scholarship. Understanding risk and motivating action to reduce it requires more than the framework at the heart of *At Risk*, the so-called "progression of vulnerability" (PAR). That framing, useful and as widely cited and applied as it has been, is limited in its assumptions about the nature of society. PAR uncritically accepts a Western worldview and is structuralist and deterministic. Although PAR values local knowledge, these limitations continue to truncate human agency. Another 20 years on, we must account for the plurality of value systems and life worlds that intersect and collide, creating both the preconditions for disasters and the abilities to prevent them (Gaillard, 2021).

A further limitation of the "progression of vulnerability" framing upon which *At Risk* depends is a failure to specify the precise kinds of power involved in driving what PAR calls "root causes" and "dynamic pressures" that combine to create unsafe conditions and fragile livelihoods. Political and economic power, as well as social, cultural, administrative, and coercive power, may be seen to be applied singly, in combinations, and sequentially (Wisner, 2022). No account of disaster risk should ignore power.

What Remains the Same

If we think again of 2004, the challenge of disaster risk was emphatically underscored by the magnitude 9.1 earthquake in the Indian Ocean and the resulting tsunami that took over 225,000 lives in a dozen countries. Despite improvements in early warning, design and engineering, and building standards, poverty and conflict continue to push people into urban areas, where they live in fragile structures in multi-hazard-prone places. Corruption and lax enforcement mean that building standards and zoning regulations may be flaunted.

Also the same in 2024 is the top-down structure of international encouragement, support, and finance for so-called "disaster risk reduction", some of which actually end up increasing disaster risk. The Sendai Framework for Disaster Risk Reduction (SFDRR) agreed upon in 2015 has taken us a step backwards from the Hyogo Framework for Action signed ten years before (Wisner, 2012, 2020). There are no additional incentives to address the root causes of risk, while the further transfer of experience and technology from wealthy to less wealthy countries is encouraged in an imperialist fashion that reminds us of the dark time of colonization. Furthermore, despite rhetorical commitment to decentralization in the so-called Grand Bargain (2016) and Grand Bargain 2.0 (2021) promoted by the UN together with non-UN agencies, local civil society organizations still receive a very small proportion of assistance without "guidance" from national or international institutions.

Why We Write

The old belief that disasters are "natural" has been overcome, and it is increasingly understood that even many hazards are caused or significantly influenced by human activity. Irrespective, the baseline is that social relations and living conditions turn those hazards into disasters. A second consensus needs to be built around the proposition that there is more than enough knowledge of the processes of disaster construction and disaster risk construction to stop it.

Enough knowledge exists to stop disasters and prevent the construction of new risks: Just apply it!

We write especially for the next generation of decision-makers, researchers, practitioners, and activists. They are the ones in universities now who will be coping with the "unresolved problems of development", as Allan Lavell defines disasters (Lavell & Maskrey, 2013), continuing conflict, and ongoing creation and exacerbation of hazards. They will hopefully shepherd in the more peaceful and life-affirming world of 2044, 20 years hence.

References Cited

Alcántara-Ayala, I., Gomez, C., Chmutina, K., van Niekerk, D., Raju, E., Marchezini, V., Cadag, J., & Gaillard, JC. (2023). *Disaster risk*. Routledge.

Blaikie, P., Cannon, T., Davis, I., & Wisner, B. (1994). *At risk: Natural hazards, people's vulnerability and disasters*. Routledge.

Gaillard, JC. (2021). *The invention of disaster*. Routledge.

Kelman, I. (2020). *Disaster by choice: How our actions turn natural hazards into catastrophes*. Oxford University Press.

Lavell, A., & Maskrey, A. (2013). *The future of disaster management*. Draft synthesis document, meeting notes, background papers and additional materials. From a scoping meeting for GAR 2015. https://www.desenredando.org/public/2013/2013-08-14_UNISDR_FLACSO_GARscoping_Document_LaRed.pdf

Marchezini, V., & Wisner, B. (2017). Challenges for vulnerability reduction in Brazil: Insights from the PAR framework. In V. Marchezini, B. Wisner, L. R. Londe, & S. M. Saito (Eds.), *Chapter 1. Reduction of vulnerability to disasters: From knowledge to action*. Rima Editora. bit.ly/3ucKyZx

Marchezini, V., Wisner, B., Londe, L. R., & Saito, S. M. (Eds.). (2017). *Reduction of vulnerability to disasters: From knowledge to action*. Rima Editora. bit.ly/3ucKyZx

UNDRR. (2020). *Time to say goodbye to "natural" disasters*. United Nations Office for Disaster Risk Reduction. https://www.preventionweb.net/drr-community-voices/time-say-goodbye-natural-disasters

Wisner, B. (2006). *Let our children teach us: A review of the role of education and knowledge in disaster reduction*. ISDR/Books for Change/ActionAid India. https://www.unisdr.org/2005/task-force/working%20groups/knowledge-education/docs/Let-our-Children-Teach-Us.pdf

Wisner, B. (2012). *What are the opportunities & challenges in the consultation process described in the background paper? Towards a post-2015 framework for DRR*. https://www.preventionweb.net/posthfa/dialogue/discussion/comment/135.html#Comment_135

Wisner, B. (2020). Five years beyond Sendai – Can we get beyond frameworks? *International Journal of Disaster Risk Science, 11*, 239–249. https://doi.org/10.1007/s13753-020-00263-0

Wisner, B. (2022). Power writ small and large. In G. Bankoff & D. Hilhorst (Eds.), *Why vulnerability still matters: The politics of disaster risk creation* (pp. 171–191). Routledge.

Wisner, B., Blaikie, P., Cannon, T., & Davis, I. (2004). *At risk: Natural hazards, people's vulnerability and disasters* (2nd ed.). Routledge.

Wisner, B., Gaillard, JC., & Kelman, I. (Eds.). (2012). *The Routledge handbook of hazards and disaster risk reduction*. Routledge.

Wisner, B., Gaillard, JC., & Kelman, I. (Eds.). (2015). *Disaster risk. Critical concepts in the environment* (4 Vols.). Routledge.

Chapter One
The Challenges of Disaster

Introduction

Disasters challenge us personally if we experience one or people we know do. We experience them socially because we empathize with those who suffer and as material or economic losses may be shared throughout the economy. We want to place disaster in a context, to frame it, and to tease out its elements and how they combine to produce injury, death, and loss. To do so, we may have to address any myths, misinformation, or lies by people with power who may want the root causes of disaster risk to remain obscure.

Among the myths that loom large in disaster studies and serve the interest of governments, researchers, consultants, and charities is that the Western approach to risk is universal or should be because it is evidence-based, scientific, rational, and therefore correct. By "Western", we mean an approach that is grounded in assertions and assumptions about reality and what it is to be human, to perceive, to think, and to reason that derive historically from pre-Socratic and classical Grecian civilization as it was transmitted and shaped by the European 17th- and 18th-century Enlightenment. We do not mean to designate a geographical region but a mentality (c.f., Bronowski, 1973; Burke, 1987). In this chapter, we show at numerous places that non-Western views and ways of knowing exist, should be respected and valued, and can provide insight. Finally, we delve even deeper into the myth of science as completely and necessarily rational and dispassionate by exploring the role of emotion in disaster studies.

DOI: 10.4324/9781003292814-1

Challenge of Understanding

"They were blind-sided!" "It was a bolt from the blue!"

There are common expressions in English that imply a belief that life can "turn on a dime", and death, injury, or loss can "reverse fortunes". English is not alone. In Spanish, there is a similar expression ("un giro de 180 grados") that suggests the need for a significant shift in approach or focus. These turns of phrase suggest that disasters may be sudden events beyond human control – accidents occurring randomly. A different view is held by many professionals and students of risk and disaster. They understand a disaster to be more than a discrete event; rather, it is the result of a long-term, slow-moving, chronic societal process of risk creation. Risk creation, or the social production of risk, is composed of decisions and actions regarding how people organize their lives with each other and society's relations with material reality (nature). At the heart of these decisions and actions is the role played by power in many forms: political, economic, social, and administrative/bureaucratic.

Consider the lowly banana peel minding its own business on the pavement. The banana peel exists as a discrete entity in my path, so I slip and crack my head. A Western systems thinker, whose ancestors (acknowledged or not) include Descartes and Enlightenment thinkers, might believe that the banana peel, pavement, shoe, foot, and head are separate things located in space-time. They believe the banana peel is a hazard (as well as enclosing a source of nourishment and pleasure: a resource as well as a resistance). For this Western thinker, there is an "event in space-time" involving these "discrete elements".

Another approach would be to consider "a situation" we could call the "banana-peel-pavement-shoe-foot-head" situation – that is, "a process", not a set of discrete things. Thus, according to this alternative view, the hazard is not an event but a process. This banana peel process sits at the intersection of many other processes, including city planning, street cleaning and sanitation, banana cultivation at a distant location, fruit importation, retail distribution, etc. Being there and slipping is part of that process. Being "there" is not a random occurrence. The Enlightenment-inspired view considers the person who slips to be a discrete individual who is "vulnerable" and who suffers the consequence of an event involving a hazard. But just as the hazard (slipped-upon-object) can be seen as a process and not a discrete event, vulnerability may be understood as "situated" and also part of a process; therefore, not a simple consequence of a short chain of cause-and-effect. The head that hits the pavement belongs to a person who is known and cherished by other human beings. The head (and person) were where they were when injured because of their relation to all those other people. The head (and person) continue to be cared for at home or, if accessible, in a hospital, where they are known as patients deserving care, even if such recognition and care is provided anonymously by caregiving human beings.

The words "process" and "situation" signal concepts that provide a vocabulary enabling an alternative "way of seeing" than the words "event" and "consequence". Event-and-consequence is part of Western common sense in daily life; whereas "process-and-situation" is not quotidian language or thinking. It is reserved for specialist and professional discourse and contemplation. For non-Western thinkers, process-and-situation belongs to a different commonsense understanding of daily life and language accessible to nonspecialists. For example, the pan-African notion of *ubuntu* asserts that people are not unique entities but only exist through and with other people. We "inter-are" as Zen Buddhist teacher Thich Nhat Hahn has written from the point of view of another non-Western culture. In the language of the Native American Hopi people, a single word is used for "white bird" and "snow flake". The word's meaning depends on season and weather conditions. Because of hundreds of years of economic dominance over much of Africa, Asia, Latin America, and the Caribbean by European powers (and internal colonialism in North America and Europe aimed at controlling Indigenous peoples), Western ontology and cosmovision have either been fully adopted or have partly modified Indigenous ways of seeing and knowing.

A minority of Western authors have departed from inherited "common sense" by asserting that "disasters are not natural". Examples are foundational pieces in a 1975 issue of *Ecologist* and in a 1976 issue of *Nature*. Books such as *At Risk*, *Interpretations of Calamity*, and *Development in Disaster-Prone Places*, along with guidelines from the same era, such as for pre-disaster and post-disaster humanitarian shelter, frame disaster as a process with numerous complex antecedents, including root causes and dynamic pressures that transmit and shape them. Such foundational publications argued that no disaster is "natural" because a group of people must be made vulnerable and exposed to a hazard before harm can arise. People are made vulnerable by the use/abuse/misuse of power by other people. Over the past few decades, power imbalances have been increasing. Human-caused climate change, species extinction, ecosystem destruction, resource overexploitation, land use transformation, and pollution result from the abuse of power.

Nevertheless, readers and writers should take caution. While a power-based, political ecological analysis of disaster risk and vulnerability is vital, it must be complemented with situated accounts of the experience of disaster. Otherwise, a structural framing without the voices of risk-bearers (all of us) as its phenomenological counterpart could reinforce silences imposed by class, race, gender, sexuality, ability, and other exclusions that an intersectional analysis can reveal.

Challenge of Loss

Cultural and Psychological Loss

There is considerable evidence of people suffering from mental hardship or trauma as a long-term consequence of experiencing a disaster in the form of post-traumatic stress disorder (PTSD) (Norris et al., 2004; Li et al., 2010; Seyedin et al., 2017). Moreover, people may grieve the loss of landscapes, an emotional state referred to as "ecological grief" (Comtesse et al., 2021) and "solastalgia" (Albrecht, 2005). Sudden, involuntary displacement from a familiar city and region, such as was experienced by tens of thousands of residents of New Orleans after Hurricane Katrina in 2005, can cause long-lasting effects on mental and physical health (Anastario et al., 2008; Fussell & Lowe, 2014; Erikson & Peek, 2022). Whilst grief at the loss of a loved one or close friend may be universal, the emotional response to the loss of a home or landmark, such as a church or temple, due to failure to prevent damage from a fire, flood, or earthquake takes on specific cultural forms, sometimes accompanied with rituals of leave-taking or letting-go, for example, in Nepal following the 2015 earthquake (Yadav et al., 2021). The importance of such culturally-specific rituals was underscored by their interruption due to the sheer magnitude of deaths in the 2010 earthquake in Haiti (Burnett, 2010).

Livelihood Loss

Livelihood loss may result from the destruction of the means of production, such as a fishing boat and nets, or reduction of access to raw materials and resources, such as die-off or migration of fish due to coral bleaching, perhaps due to natural climate variabilities, a warming climate, or a localized human act, such as cyanide fishing. Government regulations on the use of natural resources or governing conditions of work may raise the cost of livelihood activities. Middle traders may manipulate markets in ways that make a prior livelihood strategy unviable. In addition, disasters involving hazards, such as earthquakes and storms, can disrupt livelihoods. Households often have few savings or assets that can see them through a prolonged period of livelihood interruption. In some cases, distant members of extended families unaffected by a disaster may provide help, but neighbors and the wider region are likely to have suffered similar losses. Governments have variable, often limited, means of assistance at their disposal, and there may be discrimination and corruption in their distribution.

Relief assistance from government or non-governmental sources can result in dependency and fail to fully assist recipients. In some cases, the economic niche of a prior livelihood will have disappeared due to infrastructure destruction. In others, displacement and resettlement create challenges to restoring a livelihood when the raw materials, markets, or other conditions are absent in the new location. For *in situ* survivors of disaster, evidence is mixed. Van den Berg (2010) found no difference in economic mobility between people affected by Hurricane Mitch and those in unaffected areas of Nicaragua. Studies of pastoral mobility and livelihoods from many parts of the world reveal a high degree of flexibility in the face of drought or politico-administrative interference, such as the enforcement of boundaries (Mekuyie et al., 2018; Nori & Scoones, 2019).

Monetary and Material Loss

Worldwide, formal property, contents, and life insurance is uncommon. Worldwide insurance penetration in 2020 was only 7% (Statista, 2023). Informal moneylenders charging high interest rates may be available to some people attempting to rectify monetary and property loss. Remittances from relatives working abroad are another source of capital for recovery. Monetary loss most often takes the form of a tangible asset, such as a house. Money invested in an enterprise that has become nonviable or has stopped for a prolonged period in the aftermath of a disaster may also be lost.

Key Definitions

Definition and Discourse

Possibly, we should have called this section "Key *Western* Definitions". The activity of "defining" is a classic example of a Western approach to reality, rooted in Aristotle's requirement that a definition names the genus and the specific difference that sets the "defined" apart from the rest of that broader genus. A taxonomy is implied: a list of things in the world and their relationship to one another, a Western view of the world made popular in the 18th century by Carolus Linnaeus.

Both definitions and taxonomies underpin the exercise of power. The production and reproduction of policies and actions constituting disaster risk reduction are discourses that sustain power relations. Discourses contribute to solidifying, validating, and organizing knowledge into normalizing and performative narratives that pervade social life through diverse channels, including policies and actions as well as institutions, the media, architectural forms, the arts, etc. (Foucault, 1969, 1976). It is possible to study discourses on disasters and unpack the nexus between knowledge and power: how particular understandings of what we call disasters, in academia and beyond, inform specific policies and actions for reducing disaster risk. Indeed, the knowledge-practice nexus may also produce policies and actions that exacerbate risks or create new ones.

Western definitions and taxonomies inform the dominant disaster discourses – definitions and taxonomies that reflect a quest for one absolute truth. A mundane bureaucratic example is the search for universal definitions and glossaries of key concepts among stakeholders (e.g., the multiple glossaries of United Nations agencies since the 1970s). These discourses have underpinned normative policies and actions for disaster risk reduction, whether associated with technocratic and war-inspired initiatives or so-called community-based approaches (Gaillard, 2021).

An example of this quest for universal truth and standardized actions for reducing disaster risk is the research leading to the normalized recommendation to drop, cover, and hold during an earthquake – whatever the circumstances, locations, individuals, and cultures – as long as an expectation exists that the structure will remain intact. Gatekeepers, such as the United States Geological Survey, monitor, critique, and demonstrate flaws in alternative recommendations, such as to shelter in a part of a room that can provide a so-called "triangle of life" (USGS, 2022). Similarly, another US government agency advises people in coastal zones prone to tsunami to "drop, cover, hold" until it is "safe to move" and then to run uphill when they feel an earthquake (NOAA, 2015). We do not dispute this information or the recommendations. Conversely, we are grateful for them, we publicize and support them, and we follow this advice, knowing how many lives these recommendations have saved and how many more will be saved. Without in any way undermining this work, we nonetheless accept our role and obligation as researchers to constructively critique and examine, from different perspectives, what we ourselves offer and accept.

In particular, viewing the world in this way is not and cannot be universal. There are other ways of seeing and experiencing the world, hence, other discourses. We contend that multiple discourses can co-exist across various cultures and societies – and still can and should be used to save lives and help people. There are multiple interpretations of disasters rather than one unique, universal, and absolute truth. Recognizing such differences does not imply abandoning such commonly agreed Western definitions and taxonomies or the practical, life-saving instructions that derive from them. Different ways of seeing and experiencing the world may allow natural phenomena, such as flooding or earthquakes (even when modified by human activity), to be considered in various ways: as hazards, as resources, or even as retribution for social misconduct.

Whilst the terms defined later have become standard in Anglophone disaster studies, all are often problematic in other languages. Consider Swahili, a language widely spoken in central and eastern Africa. "Hazard" is most often rendered as *hatari*, a word that also means "danger". *Kuwemo hatarini* means "to be in danger". This certainly captures one aspect of hazardousness, but it lacks the resonance of the binary "resource-hazard" that is common in Anglophone geographical discourse. The binary in English suggests that the material environment is Janus-faced and can present as something beneficial or harmful, depending on human actions. More serious problems in translation arise concerning the rest of these standard English terms. "Vulnerability" glosses as *kuathirika*, a word that means "affected". *Kuathiri* means "to affect", and *athiri* translates as "an effect". The Swahili does not capture the multiple aspects of "vulnerability" – including but not limited to exposure and sensitivity to loss or harm. "Coping" can be rendered as *kukabiliana*, but this simply means to counter or to push back against, to resist. Again, this is just one of the aspects of coping, which involves means, method, and resources, hence, capacity and capability. These latter two are both translated as *uwezo* – ability. Finally, "risk" is the most troublesome term. Swahili uses the same word used for risk and hazard: *hatari*. Disaster studies from the 1970s onward has rejected the identity of hazard and risk (Wisner et al., 1977), having rejected an older technocratic paradigm that did identify the two (and, hence, naturalized risk). The conventional understanding today is that risk is a function of hazard and vulnerability (often with other words, such as exposure and capacity to cope, involving a broad range of capabilities). It is not possible to express that nuanced and complex notion to a monolingual Swahili speaker without circumlocution, further explanation, and perhaps examples.

Hazard

Consider the spinning sphere (oblate spheroid or ellipsoid), the surface of which is our home. It rotates around a star and is tilted such that energy from that star is unevenly spread on the surface, giving us seasons. Its molten core moves. The relatively thin crust is broken into plates that move several centimeters or more each year. They ride up over one another and slide down below. Mountains rise slowly. Rain erodes them and meanders over floodplains. The dynamism of such a world is something most take for granted – seasons, the hydrological cycle, distribution of vegetation, other living beings, and non-living beings.

Human beings have adapted to all this reasonably well. Humans have survived and prospered for perhaps up to 2.8 million years, if we take the earliest suggested emergence of the genus *Homo* (Prat, 2022). Yet each of the geophysical, geomorphological, and hydrological processes that have shaped the Earth is capable of releasing energy and producing forces in great excess than that to which we are accustomed. The river that provides water, fertile alluvium, and fish may flood. The volcano whose slopes are covered with fertile soil derived from ash may erupt. The faults in the rock that provides upwelling water may move and destroy dwellings. The restless planet is a Janus-faced home: sustainer and destroyer.

The hazards humans face vary in space and time. For example, epidemiology and medical geography have studied the distribution of health hazards in space and time at various scales from the urban neighborhood to biomes with diurnal, seasonal, annual, and other periodicities. In the U.S., the chance of vehicle crashes is highest in the evening and during the rush hour (National Safety Council, 2022). Exposure to mosquito-borne parasites and viruses tends to be greater in tropical than subtropical latitudes.

Risk

To exist and to be conscious of one's own existence is rare. According to a Buddhist parable, to be born a human is as likely as a sea turtle rising out of the ocean in a manner that traps its head in a wooden yoke that happens to be floating there (Barnett, 2019). Rare as it is, human life (indeed, all biological life) can be ended at any time due to various forms of harm, such as physical trauma or disease. Life is contingent, a light that illuminates a small corner of darkness. In the context of such contingency, fragility, and precarity, it is not surprising that daily life is risky. Some privileged people have the leisure to indulge in ignorance of this fact of life (and death). Less privileged individuals intuitively understand risk as the chance that something bad or harmful may happen.

What people perceive as a risk or as risky is influenced by culture, experience, personality, age, gender, and many other factors. Risks chosen for preventive or mitigative actions and the methods chosen to reduce risk are largely determined by culturally diverse perceptions and understandings (Alaszewski, 2015). Open-ended discussions reveal that culture and situation may give rise to risk priorities that seem counterintuitive to those outside the particular culture and situation. For example, lack of easily accessible clean freshwater may trump landslide dangers in a mountainous situation, while concern with conflict may spontaneously arise from people who appear to outsiders to be vulnerable to flooding (Gibson & Wisner, 2016). Box 1.1 on the COVID Tracking Project provides an example of a successful local discussion.

BOX 1.1 The COVID Tracking Project

The COVID Tracking Project is an excellent example of citizen science and the power of local knowledge. It began in early March 2020, when two journalists at *The Atlantic* magazine, Robinson Meyer and Alexis Madrigal, built a tracker to investigate lagging COVID-19 testing rates. Jeff Hammerbacher had a separate tracking spreadsheet. The two efforts came together on 7 March, calling for volunteers.

A managing editor named Erin Kissane joined and the COVID Tracking Project was born. The project grew to around 30 paid staff and 250–300 volunteers. Data were collected for one year. Every day, during the year covered by the tracking project, data was collected by a network of volunteers from all 50 US states, five territories, and the District of Columbia (https://covidtracking.com/data). This dataset has been used by national and local news organizations across the US and by research projects and agencies worldwide. Donated software allowed sites and apps to import the dataset automatically as volunteers, advisors, and editors were feeding into it. During that period, the project received about 2 million requests per day.

Meanwhile, on 15 April 2020, the project launched an additional initiative, the COVID Racial Data Tracker (https://covidtracking.com/race), a partnership between the COVID Tracking Project and the Center for Antiracist Research based at Boston University (https://www.bu.edu/antiracism-center/). The project collected, analyzed, and published racial data on the pandemic within the US until March 2021 (https://www.bu.edu/antiracism-center/antiracism-research/racial-data-lab/). It also collected COVID stories and narratives by people of color (https://www.bu.edu/antiracism-center/antiracism-research/covid-stories/).

Source: The COVID Tracking Project (https://covidtracking.com/about)

Vulnerability

Vulnerability, too, is complex because of its variations in space and time and also because it involves many characteristics of people and groups that expose them to harm and limit their ability to anticipate, cope with, and recover from harm (Wisner et al., 2004). Disciplines have different ways of defining, measuring, assessing, and interpreting vulnerability. These include public health, psychology, law, geography, political science, engineering, medicine, and development studies, among others. Some practitioners focus on the short-term identification of vulnerability so that maps and lists of people "living at risk" can be created and used by authorities to intervene, control, regulate, and discipline. Other professionals and activists are more concerned with the reasons why some people are more vulnerable than others when facing a threat or hazard. Those working at the scale of localities may be interested in methods that bring out residents' own knowledge and experience of hazards and may help people to cooperate with one another to find ways of reducing risk (Gibson & Wisner, 2016). There are some interpretations of vulnerability that seek its root causes and inquire into the process of risk creation by the routine function of political (including economic) systems in which small elite groups determine locational and investment decisions for their own benefit without regard to how these decisions affect the majority. Finally, whatever success there may be in attempts to reduce it, vulnerability remains part of the human condition, and this fact is both haunting and puzzling (Wisner, 2016b; Armstrong, 2017).

Coping, Capacity, and Capability

These three words are often used interchangeably. Coping more often refers to real-time, short-term actions taken to avoid or to reduce harm in the face of a hazard (Wisner et al., 2004). It carries with it the sense of spontaneous, one-off actions that are not a part of a plan or strategy, although it may be influenced by experiences in the past. Capacity has been used to signify specific latent skills and characteristics that enable people to cope with and recover from disasters; whilst capability (or sometimes simply "ability") has a broader scope that takes in the social, including political-administrative (which incorporates economic) arrangements that allow people access to education, information, and other resources required for effective coping, development of capacities, and abilities (Nussbaum, 2011; Gaillard, 2014; Robeyns & Byskov, 2020).

Mitigation of Risk

As discussed earlier, individuals may use abilities that allow them to resist, cope with, or modify hazards and to reduce vulnerability (including exposure) and risk. Such action can be called "personal protection". Collectivities, institutions, and organizations, including governments at different scales, attempt to provide "social protection". Social protection includes measures that exceed a narrow focus on disaster risk and hazards. It can embrace a broad range of entitlements and institutional arrangements that are believed to promote as well as protect health and well-being. We mention this all-embracing sense of social protection in order to highlight the importance for human well-being of good health and nutrition, education, social connection, means of livelihood, security at all ages, including in retirement, and access to public decision-making. People who enjoy such a situation are less vulnerable to life disruptions, including hazard impacts, and are more robust in their ability to recover and adjust their lives.

More narrowly focusing on disaster risk, governmental and non-governmental actors have a range of policy tools and practices available, depending on resources. They may attempt to affect the hazard itself, for example, by engineering works, such as dams, levees, and dikes. Exposure may be the focus of public policy through zoning regulations and their monitoring and enforcement. The state may expend resources on public education concerning hazards and personal protection as well as on warning systems. Building codes are another tool for risk reduction. As numerous earthquakes around the world show, monitoring and enforcement cannot be assumed to be universal and thorough (Lewis, 2003). Lewis (2003) showed that corruption is as large a hazard as anything that our turbulent and restless Earth might throw at us. Twenty years after Lewis (2003) drew this conclusion, his explanation was borne out afresh by the earthquakes of 6 February 2023 along the Türkiye-Syria border.

Disaster risk can be ill-governed by public authorities lacking the knowledge and skill or the resources to make informed decisions or implement policies. Disaster risk can be misgoverned by officials who make decisions and push forward policies that favor a minority elite and possibly outside investors without regard for the impacts on the majority. In some cases, ethnic minority groups who face hazards, such as drought and flooding in peripheral regions of a country, may be neglected by metropolitan central government, giving rise to marginal people living in marginal places (Wisner, 1978; 2016a; Walters & Gaillard, 2014). This, too, is a form of misgovernance. Finally, risk mitigation may involve well-meaning political leaders and government officials in moral hazard. For example, the evacuation ordered by authorities along the US Gulf Coast in anticipation of Hurricane Rita in 2005 caused many more deaths from vehicle crashes and from heat stress in traffic jams than the storm itself, although we do not know how many would have died in the storm without the evacuation (Zachiria & Patel, 2006).

There is a common expectation among citizens of contemporary Western democracies that the state has a responsibility to protect by providing the means of social protection. Since at least the 17th century, this attitude has been assumed as part of the social contract between the citizen and the Leviathan state that has a monopoly over the means for doing so on a large scale and the means of destruction (Hobbes, 1651). Whilst there is some truth in this assumption, the state may use disaster risk reduction measures to control and discipline citizens (Gaillard, 2021) and to enlarge its power. Indeed, some have argued that disaster management was the engine of early modern state creation (Bandopadhyay, 2022).

The Western View of Risk

The Western approach to disaster blends the key elements just defined into a variety of frameworks where they serve as mnemonic devices ensuring that useful questions about disaster risk are not forgotten and that a practitioner, researcher, or activist is nudged to consider asking them. A number of these frameworks are discussed in Chapter 2. To give just one example, consider the following:

$$R = (H \times V/C) - M$$

Where R represents risk, H stands for hazard, V/C is the ratio of vulnerability and capacity to cope or individual protection, and M stands for mitigation of risk (hazard management and/or social protection). We must repeat that this is a mnemonic device, not a real equation, the terms of which would stand for variables or parameters whose values can be specified and always quantified. Indeed, vulnerability, capacity, and mitigation are matters of social relations and power. They are qualitative relations, not quantities and not fully quantifiable. So, ultimately, is risk, although many people try, and there is often confusion between technical indicators and political power. Within the context of "normal science" and Western common sense, this mnemonic seems innocent of bias or cultural baggage. It is filled with assumptions about individualism, the role and power of the state, the separation of humans from other biological beings, and the capacity of humans to exert control over material reality and their own lives.

Origin

Until the 1970s, a single dominant Western discourse identified disaster with the hazard that accompanies it. The dominant view and a contrast are evident in a famous letter from Rousseau to Voltaire (Rousseau, 1756; Dynes, 2000), questioning the 1755 Portugal earthquake as the disaster by indicating that city construction and people's behavior caused the witnessed problems, whereas an earthquake far from habitation might not lead to destruction. Following Rousseau's sentiments and logic, scholars in the 17th and 18th centuries thought that science was adequate to eliminate "accidental" death and free humanity from suffering (Dekker, 2017) and that reason, scientific understanding, development of technology, and perfection of political order were how human life could improve (Bakewell, 2023). Also from the late 1600s, modern forms of property insurance became available – for example, in London after the Great Fire of 1666.

These Enlightenment centuries saw the notion of randomness become associated with risk as the ancient association of risk with fate fell away. In 1620, Galileo published "On a discovery concerning dice" that took a mathematical approach and introduced the role of randomness and probability. A probabilistic approach to risk implies that the utopian "zero vision" (no accidents and, by extension, no disasters) is impossible. The random combination of complex events will always produce a nonzero probability of loss, injury, or death (Moses, 2018). Thus, hidden away beneath the rationalistic surface of the self-confident, Enlightenment "zero vision" was still the "bad star" (etymological origin of "dis-aster" and, therefore, a residue of astrology).

The second Western discourse is that of risk management and risk reduction. A shift was made from focusing on the hazard to placing emphasis on exposure to hazard and the vulnerability of affected people, livelihoods, and assets. Individuals and small self-organized groups may or may not be able to avoid or absorb the losses due to hazard impacts. The Western discourse focuses on the institutions of the state as the main source of risk management. In some versions of this discourse, a whiff of Victorian moralism remains in an appeal to individual responsibility to be aware of hazards and to follow authoritative guidance. More critical views inquire into the role of the state and other holders of power as creators of risk, yet to date, the study of disaster risk creation lags behind mainstream disaster risk reduction scholarship (Lewis & Kelman, 2012; Wisner, 2020).

Dissemination and Consequences

Disaster reveals human vulnerability while calling forth human potential to rearrange daily life, communities, and nature so that disaster risk can be reduced. This is the heart of the humanist interpretation of the 1755 Portugal earthquake: a belief in progress and betterment that has been repeated continuously since and disseminated around the world by 19th-century colonialism and subsequent neocolonial and postcolonial regimes of economic power. The discourse of vulnerability became prominent again during the Cold War and was internationalized by the combined forces of economic globalization and neoliberal policy enforced by the World Bank and the International Monetary Fund (Bankoff, 2020). "Resilience" and "adaptation" joined vulnerability as key tropes after the fall of the Soviet Union in 1991. Bankoff (2020, p. 27) observes that "no rival discourse seems ready yet to challenge Western hegemony in the language and metaphor of international governance and development policy". He continues, "[T]he power relations that underlie these discourses have not changed significantly since the Second World War, as the world has largely remade again and again according to an image fashioned by certain sectors in Western societies".

A consequence of near-universal hegemony of Enlightenment-inspired disaster management is to provide the owners of capital carte blanche to "tame" nature through environmental management while investing in the exploitation of mineral and energy sources worldwide, as well as to create ever more elaborate value chains of food, fiber, and even wild genes (Shiva, 1999, 2020). Another consequence is to enshrine the state as the protector of people who fail in their responsibility to protect themselves from hazards – ironically, even those who are harmed by capital's rapacious quest for resources and the pushback by nature at attempts to tame and control (think of the emergence of new viruses from forest and plantation frontiers) (Wallace, 2016).

Multicultural Understanding(s)

The Western framework is used by professionals all over the world due to political (including economic) globalization and neocolonial dependency (Blaut, 1993). Side by side, people who are these professionals' compatriots are likely to entertain cultural differences in how they understand risk and disaster. This is one reason why there is often incomprehension and sometimes mistrust between Westernized disaster managers and the majority of the population. Consider the following examples from around the world.

Africa

Africa is home to more than 1.2 billion people speaking more than a thousand languages. For the sake of illustration, we will consider a single language and its speakers in eastern and central Africa: those speaking Swahili. These remarks require caution because most Swahili speakers grew up speaking local languages (perhaps 126 in Tanzania and 68 in Kenya). Assuming that language influences perception and understanding (Sapir, 1921; Whorf, 2012), the combination of vernacular, Swahili, and for educated and urban people, English, may produce differences in how people process their reality.

As mentioned earlier, Swahili does not differentiate between risk (*hatari*) and hazard (*hatari*). The word for catastrophe, plague, and epidemic are the same (*janga*), whilst the common word used in the Swahili-language media for pandemic (frequently used for COVID-19) is *janga kubwa* (large epidemic). The English "disaster" may be rendered as *maafa*, with the caveat that for native speakers and Afrocentric scholars elsewhere, this Swahili word brings to mind the devastation of the slave trade. This Swahili term was repurposed to stand for the slave trade by Marimba (1988), where it also has the meaning of "disaster, terrible occurrence or great tragedy".

As many as a third of the words in Swahili have an Arabic root, and some Tanzanians and Kenyans living on the Indian Ocean coast, especially Muslims who have attended Koranic schools, know some Arabic. This introduces another possible source of ambiguity. The Arabic for "disaster" (*karitha*, كارثة) is but one of at least eight terms for calamity, misfortune, scourge, ordeal, distress, tribulation, and related terms.

In addition, Swahili speakers and other Bantu peoples share a common core understanding of illness and other misfortune as originating in social conflict and rupture of harmonious family and community relations. For the vernacular Swahili speaker and many other Africans, the argument that disaster risk is socially constructed does not need to be made. They understand the origin of misfortune to be located in broken relationships and a rupture of harmony.

Asia and the Pacific

The Austronesian world that spans from Madagascar and Southeast Asia to Melanesia and Polynesia is home to more than a thousand languages. To illustrate, we focus here on a single tongue and its speakers: the Kapampangans of Central Luzon in the Philippines. Very little is known of precolonial time except for sparse archaeological records and some early accounts by Spanish chroniclers in the late 16th century that were inherently filtered by a colonial lens. As such, the contemporary contours of the Kapampangan identity have largely been shaped by colonial referents, which include using Latin grammar to formalize its language, Western events as reference events in Kapampangan history, and Christianity as the dominant religion (Gaillard, 2023). The Kapampangan identity is nowadays a hybrid that combines some elements of precolonial culture, including influences from other Asian societies, with a Western colonial legacy. The hybrid nature of the Kapampangan identity is most evident in its language.

Interestingly, there are two words most often used today by Kapampangan speakers to translate the Western meaning of "disaster": one is *kalamidad*, a loan from Spanish (*calamidad*), and the other one, most often considered Indigenous but which is not, is *sakuna*, a loan from Sanskrit (शकुन), where it refers to a *bird of omen*. The latter is particularly interesting because in Bahasa Indonesian, the word most often used for disaster is *bencana*, another Sanskrit loan (वञ्चन) that means trickery. Even more interesting is that in Hindi today, there are other Sanskrit terms, *aapada* (आपदा) and *vipatti* (विपत्ति), used to capture what the West means by disaster. *Kalamidad* and *sakuna* are used interchangeably in everyday Kapampangan language today, the latter probably more often than the former. *Bencana* is the official Bahasa Indonesian word for disaster promoted by the government. It is used widely by academics and the media in Indonesia.

In the Philippines, the lack of an Indigenous Austronesian word for the Spanish *calamidad* and English "disaster" puzzled both Spanish and American colonizers. The Spanish friars and American teachers who studied the Kapampangan language looked for what they thought would be the closest equivalent. They settled on *tagku*, for which the closest translation is misfortune, which was found in an 18th-century Spanish dictionary (Bergaño, 2007) and eventually in an early 20th-century American glossary (Parker, 1905). The use of *tagku* as an equivalent for *calamidad*/"disaster" never gained traction among local people, so it has disappeared from contemporary vocabulary. The continuing use of outside loan words in contemporary Kapampanga ultimately demonstrates the following:

- The Western approach to understanding harm, hardship, and suffering did not exist prior to precolonial trading relations with India and colonial contact with the Spanish conquistadores in the late 16th century.
- Spanish and, eventually, American colonizers forced their own interpretation of the world, although with limited success, upon the local people.

The Americas

We look at Latin American Spanish(es) and Brazilian Portuguese while considering the many languages spoken by Indigenous peoples, such as Diné, Quechua, Guaraní, Aymara, Nahuatl, and versions of Mayan alongside dialects such as Haitian Creole.

In Spanish, the words *catástrofe* (catastrophe) and *desastre* (disaster) are often used interchangeably, although the former is the older term. The word "catastrophe" comes from ancient Greek, which means "against" and "to become", while "disaster" comes from the Latin words "des" (negative, unfortunate) and "astre" (astro, star). The latter expresses an unfortunate occurrence caused by the stars or gods beyond human control. The Royal Spanish Academy (RAE) Dictionary distinguishes between "disaster" and "catastrophe". It defines "disaster" as a great misfortune or unfortunate event and a "bad result or bad organization". "Catastrophe" refers to an unfortunate event that disrupts the regular order of things and a sudden change of state in a dynamic system caused by a slight alteration of one of its parameters. The RAE Dictionary defines "risk" as *riesgo*, which comes from the old word *riesco* because of the danger it poses. "Contingency" or "proximity of damage" and *peligro* (danger or threat) come from the Latin word *pericŭlum*, which refers to an imminent risk or contingency of something bad happening. *Peligro* also refers to a place, step, obstacle, or situation where the imminence of damage increases. When someone is "in danger", they are being exposed to a risky situation, in contrast to interpretations and understandings in the Aztec culture (Box 1.2) (Figure 1.1).

BOX 1.2 Tetzáhuitl: Omen and Fate in the Aztec Culture

FIGURE 1.1 The fourth sign of the coming of the Spanish: the comet

Source: Public domain. Codex Duran, page 1

The term "disaster" is not embedded in Aztec culture. Instead, there is the notion of signs or omens foretelling fate. The word Tetzáhuitl is used, and its cosmological meaning is related to the consequences of disastrous events of a diverse nature. A Tetzáhuitl breaks the usual and daily order of the world. It is something anomalous or unusual that manifests itself in order to announce and provoke generally negative future events. Augury was practiced by Aztec priests. With such occult knowledge and practice, a sign can reveal to human beings the knowledge of a divine determination that has occurred but not yet been carried out.

The gods informed humans of great future catastrophes through signs of an extraordinary nature and magnitude. These catastrophes included the destruction of kingdoms, famines, epidemics, wars, and the deaths of rulers, along with the Spanish invasion and conquest. Warnings took the form of unusual celestial events, such as comets and lights, and the appearance of monstrous beings (for example, a crane with a mirror on its head and two-headed men), "accidents" that seemed to have an insufficient physical cause, such as temple fires, waves without wind, and lightning in calm weather.

Fate was considered a divine action. In cosmological terms, it was the occasion when a god penetrated the world and acted in it. So the Tetzáhuitl, or omen, is the manifestation of a divine will that has already arrived in the world but that has not yet been fully realized and which can be perceived by human beings through effects on other creatures or circumstances that serve as signals. If the gods produce – or at least allow – such warnings, then it can be inferred that they also allow human beings to intervene in their own future. The further implication is that human action can effectively modify or avoid the circumstances about which they have been warned.

Sources: López-Austin (2019); Sahagún (1979)

Brazilian Portuguese has its terms for catastrophe (*catástrofe*), disaster (*desastre*), calamity (*calamidade*), emergency (*emergência*), and tragedy (*tragédia*), among others. *Risco* is the equivalent of risk, while *vulnerabilidade* is the Portuguese translation for vulnerability. But there are some scholars and activists who use *vulnerabilização* to refer to social processes that create the conditions of vulnerability. More challenging is the translation of the word "hazard". The Portuguese word *azar* means "bad luck". The pronunciation of *azar* is very similar to "hazard". Scholars and civil defense officers translate "hazard" as *ameaça* (equivalent to "threat") or *perigo* (equivalent to "danger"). People living in risk-prone areas more often use *perigo* or refer to the specific phenomenon such as flood (*inundação*) or landslide (*deslizamento, escorregamento* or *queda de barreira*).

There are regional variations for terms such as "floods". Most states use *inundação* (riverine floods), *enchente* (flooding), *alagamento* (urban flooding), or *enxurrada* (flash floods) as synonymous. In Acre in the Brazilian Amazon, people use the term *alagação* to refer to flood. Other states in the Brazilian Amazon use the word *cheia* to refer to flood.

The Diné (Navajo) people of North America understand illness as disharmony with the universe and view hazards, such as wildfires, as the result of disharmonies among people and between people and the land. During interviews with Diné elders, Redsteer et al. (2015) were told that an historic shift away from traditional grazing practices has produced more disputes among individuals who claim that permits give them exclusive grazing rights. Settlement patterns have been affected, and close kin are no longer near enough to cooperate in ceremonial activities. Further, they describe a shift away from traditional Diné cosmological knowledge and practices. They suggest that people who no longer depend on livestock become insensitive to the needs of the land. Earlier, their elders offered prayers to ancestral spirits for the blessing of rain and abundant vegetation.

In an interview with Avery Denny, a traditional Diné practitioner and faculty member at Diné College, journalist Alaistair Lee Bitsoi wrote in *Diné bi Naltsoos* (*Navajo Times*) (Bitsoi, 2011):

> Traditional Navajo medicine people view the world's most recent natural disasters, including the human-caused Wallow Fire and other raging wildland fires throughout the U.S., as Mother Nature's warnings of the imbalanced state of the natural world.

Denny was reported as saying that the exploitation of natural resources is the main reason for the natural world's imbalance, and he continued:

> In the natural world, we have the Holy People, . . . The four elements of life - fire, air, water and earth are the *Diyin Dine'é*. We have to learn to respect them or else a natural disaster occurs.

Challenge of Contention

Why "Frame" Disaster?

One can think of a framework as a machine for generating useful questions. "Useful" here means relevant to action for the benefit of living and non-living beings. When one encounters situations involving people and material reality (say a street festival or a landslide that has buried houses on a similar street), emotions arise. In the first case, perhaps sympathetic joy arises. In the second, perhaps sorrow and empathy for suffering and loss. Emotion is followed rapidly by these questions: How am I involved? Should I dance with the merrymakers? Should I search for injured people? All sorts of other questions may arise along the lines of what, where, when, who, and why? "Why" may come later if one is immediately involved in the landslide and rescue operations. A framework is a way of formalizing and prioritizing all these questions in a context. The framework of default for the geomorphologist privileges the questions "what" and "where", in addition to the physical processes of "why". The framework most natural for the trained responder lays its initial emphasis on the questions "who" and "where". For the critical social scientist, the fundamental question is "why". It is the question that gives rise to inquiry into disaster risk creation.

What Is Nature? Why Do Humans Think They Are Separate?

Human bodies, including their brains, comprise the same elements as other living beings around them – carbon, nitrogen, oxygen, calcium, and phosphorus, among others – leading to some similar dominant molecules, such as water. We share our physical being with all of nature, including non-living beings. Yet the Oxford English Dictionary (2006) defines "nature" as "the phenomena of the physical world collectively, including plants, animals, the landscape, and other features and products of the earth, *as opposed to humans or human creations* [emphasis ours]; the physical force regarded as causing and regulating these phenomena". Few people would disagree. It is presumed to be common sense that we humans are separate from the rest of living beings and material reality, including non-living beings. Our separateness could well be a delusion brought about by our cravings and aversions (desire for pleasure and fear of pain) that motivates actions affecting nature (material reality). These actions allegedly once included transporting lightning-kindled fire into our dwelling and nurturing that flame, learning its uses. Over hundreds of thousands of years since, the actions of humanity have accumulated and interacted in such a way that much of the surface of the planet has been modified, many species have been driven to extinction, plants and animals have been domesticated, hybrid species have been created, and the atmosphere, waters, lands, and livingness of our world have changed. All of this busy activity has cultivated the understanding of human separateness.

Despite the shift in attention from the time of Rousseau's (1756) letter to Voltaire, from a focus on hazard to concern with vulnerability, the nature/hazard versus society/vulnerability binaries persist. Human claims of separateness persist. Our attention and paradigms have changed radically. Some no longer blame deities or nature for disaster, although some do, but irrespective, we are still stuck with an understanding of disaster at the interface of nature and society, a problematic binary inherited from the Enlightenment (Castree & Braun, 2006).

Disasters Are Not Natural

A major watershed in the understanding of disasters took place in the 1970s and 1980s, when it became widely accepted that disasters are not natural (Ball, 1975; O'Keefe et al., 1976). Humans and other living beings are exposed to potentially harmful processes, such as landslides, floods, and wildfires. We are vulnerable to harm, but in the case of humans, we are not all equally vulnerable. Our social nature (sometimes referred to as our second nature since we are, as we have seen, made up of natural elements) has given rise to power relationships among individuals and groups, as well as between groups. These power relations affect where and when individuals and groups reside and where they pursue their livelihoods, spaces through which they move, and how resources are distributed. Power relations further affect what, when, and who enjoys the possibilities for nourishment and pleasure in a place and who is exposed to the potential harms or hazards of that place.

Disasters Are Complex – and Simple

Disasters are not natural, but hazards sometimes are. People are not innocent victims of natural hazards, but as we have seen, people are busily active in shaping nature and in affecting one another. There are no innocent disasters. People may act so that other people are placed in harm's way. People may act on the material surrounding them in ways that exacerbate hazardous conditions. These sorts of human actions are part of deeply rooted social patterns, the same patterns that gave rise generations ago to caste and class societies. Therefore, disasters result from long-term, slow-moving societal processes that also shape nature – a simple explanation with complex occasionings, consequences, and responses (Box 1.3).

BOX 1.3 Aggregate, Compound, and Cascading Hazards

A combined Web of Science and Scopus review in 2022 of 91 papers revealed three different kinds of interaction among hazards (Drakes & Tate, 2022): aggregate, cascading, and compound. Aggregate hazards do not interact as such; however, their impacts are additive. Cascading hazards may affect each other and produce a further hazard, all of whose impacts are also additive. For example, a hurricane disrupts electricity supply that in turn results in a chemical spill. Compound hazards may have completely different causal pathways and spatial-temporal scales but nonetheless result in significant impacts. An example is people sheltering from the explosive eruption of a volcano of St. Vincent in the Caribbean in 2021 during the COVID-19 pandemic and a dengue fever outbreak, with shelters arranged to be at approximately half of their typical numbers (Ferdinand & Badenock, 2023). Compound and cascading impacts were shown in an empirical study of German farming and livestock management during the drought of 2018–2019: crop losses were usually followed up by a shortage of feed for livestock and consequent early slaughtering of animals (De Brito, 2021).

Conclusions

All the preceding demonstrates that disasters are not discrete events but rather processes of overlapping contingency taking place in daily life and occasioned by societal processes. Note the phrase "occasioned by" and not "caused by". The mechanistic understanding of causation is, in some understandings, that one discrete event precedes and provides both necessary and sufficient conditions for a subsequent event to occur. This approach is frequently seen, for instance, in theories of "cascading disasters". For disasters, necessary conditions are complex and contingent, involving the world (material reality) and people's actions. Understanding disasters is not simple nor is it easy. As we conclude Chapter 1, we hope that we have made a "useful" start.

References Cited

Alaszewski, A. (2015). Anthropology and risk: Insights into uncertainty, risk and blame from other cultures – a review essay. *Health, Risk and Society, 17*, 3–4. https://www.tandfonline.com/doi/full/10.1080/13698575.2015.1070128

Albrecht, G. A. (2005). 'Solastalgia'. A new concept in health and identity. *PAN: Philosophy Activism Nature, 3*, 41–55.

Anastario, M., Larrance, R., & Lawry, L. (2008). Using mental health indicators to identify postdisaster gender-based violence among women displaced by hurricane Katrina.

Journal of Women's Health, 17(9), 1437–1444. https://www.liebertpub.com/action/doSearch?SeriesKey=jwh&AllField=Hurricane+Katrina

Armstrong, F. (2017). An extrinsic dispositional account of vulnerability. *Les ateliers de l'éthique. The Ethics Forum, 12*(2–3), 180–204. https://doi.org/10.7202/1051281ar

Bakewell, S. (2023). *Humanly possible: Seven hundred years of humanist freethinking, inquiry, and hope*. Penguin.

Ball, N. (1975). The myth of the natural disaster. *The Ecologist, 5*(10), 368–369.

Bandopadhyay, S. (2022). *All is well: Catastrophe and the making of the normal state*. Oxford University Press.

Bankoff, G. (2020). Remaking the world in our own image: Vulnerability, resilience, and adaptation as historical discourses. In G. Bankoff & D. Hilhorst (Eds.), *Why vulnerability still matters: The politics of disaster risk creation* (pp. 15–32). Routledge.

Barnett, B. (2019, December 8). *More yokes for more folks. The Lion's roar: Buddhist wisdom for our time*. https://www.lionsroar.com/more-yokes-for-more-folks/

Bergaño, D. (2007). *Vocabulary of the Kapampangan language in Spanish and dictionary of the Spanish language in Kapampangan: The English translation of the Kapampangan-Spanish dictionary* (p. 519). The Juan D. Nepomuceno Center for Kapampangan Studies Holy Angel University.

Bitsoi, A. (2011, June 23). Disasters signal imbalance in the natural world. *Diné bi Naltsoos (Navajo Times)*. https://www.navajotimes.com/news/2011/0611/062311tradition.php

Blaut, J. (1993). *The Colonizer's model of the world*. Guilford.

Bronowski, J. (1973). *The ascent of man*. Little and Brown Company.

Burke, J. (1987). *The day the universe changed*. Little and Brown Company.

Burnett, J. (2010, February 3). Earthquake takes its toll on Haiti's burial rites. *All Things Considered (Rational Public Radio)*. https://www.npr.org/templates/story/story.php?storyId=123331644

Castree, N., & Braun, B. (2006). Constructing rural natures. In P. Cloke, T. Marsden, & P. H. Mooney (Eds.), *Handbook of rural studies* (pp. 161–170). Sage.

Comtesse, H., Ertl, V., Hengst, S., Rosner, R., & Smid, G. (2021). Ecological grief as a response to environmental change: A mental health risk or functional response? *International Journal of Research in Public Health, 18*(2), 734. https://doi.org/10.3390/ijerph18020734

De Brito, M. M. (2021). Compound and cascading drought impacts do not happen by chance: A proposal to quantify their relationships. *Science of the Total Environment, 778*, 146236. https://doi.org/10.1016/j.scitotenv.2021.146236

Dekker, S. (2017). Zero vision: Enlightenment and new religion. *Policy and Practice in Health and Safety, 15*(2), 101–107. https://doi.org/10.1080/14773996.2017.1314070

Drakes, O., & Tate, E. (2022). Social vulnerability in a multi-hazard context: A systematic review. *Environmental Research Letters, 17*(3), 033001. https://doi.org/10.1088/1748-9326/ac5140

Dynes, R. (2000). The dialogue between Voltaire and Rousseau on the Lisbon earthquake: The emergence of a social science view. *International Journal of Mass Emergencies and Disasters, 18*(1), 97–115. https://doi.org/10.1177/028072700001800106

Erikson, K., & Peek, L. (2022). *The continuing storm: Learning from Katrina*. University of Texas Press.

Ferdinand, I., & Badenock, K. (2023). Emergency shelters fleeing La Soufrière: A case study from the April 2021 volcanic eruption in St Vincent and the Grenadines. *Geological Society, London, Special Publications, 539*, 365–392.

Foucault, M. (1969). *L'archéologie du savoir*. Editions Gallimard.

Foucault, M. (1976). *L'Histoire de la sexualité*. Editions Gallimard.

Fussell, E., & Lowe, S. (2014). The impact of housing displacement on the mental health of low-income parents after hurricane Katrina. *Social Science and Medicine, 113*, 137–144. https://doi.org/10.1016/j.socscimed.2014.05.025

Gaillard, JC. (2014). *People's response to disasters in the Philippines: Vulnerability, capacities and resilience*. Palgrave-Macmillan.

Gaillard, JC. (2021). *The invention of disaster: Power and knowledge in discourses on hazard and vulnerability*. Routledge.

Gaillard, JC. (2023). *Lakbe: A collection of essays in contribution to Kapampangan studies*. Center for Kapampangan Studies.

Gibson, T., & Wisner, B. (2016). "Let's talk about you . . .": Opening space for local experience, action and learning in disaster risk reduction. *Disaster Prevention and Management, 25*(5), 664–684.

Hobbes, T. (2017). *The Leviathan*. Penguin Classics. (Original work published 1651)

Lewis, J. (2003). Housing construction in earthquake-prone places: Perspectives, priorities and projections for development. *Australian Journal of Emergency Management, 18*(2), 35–44.

Lewis, J., & Kelman, I. (2012). The good, the bad, and the ugly: Disaster risk reduction (DRR) versus disaster risk creation (DRC). *PLOS Currents, 4*, e4f8d4eaec6af8. https://doi.org/10.1371/4f8d4eaec6af8

Li, X., Huang, X., Tan, H., Liu, A., Zhou, J., & Yang, T. (2010). A study on the relationship between posttraumatic stress disorder in flood victim parents and children in Hunan, China. *The Australian and New Zealand Journal of Psychiatry, 44*(6), 543–550. https://doi.org/10.3109/00048671003601400

López-Austin, A. (2019). Las señales. La palabra tetzáhuitl y su significado cosmológico. *Estudios De Cultura Náhuatl, 57*, 13–29. https://nahuatl.historicas.unam.mx/index.php/ecn/article/view/77873

Marimba, A. (1988). *Let the circle be unbroken: The implications of African spirituality in the diaspora*. Nkonimfo Publications.

Mekuyie, M., Jordaan, A., & Melka, Y. (2018). Understanding resilience of pastoralists to climate change and variability in the Southern Afar region, Ethiopia. *Climate Risk Management, 20*, 64–77. https://doi.org/10.1016/j.crm.2018.02.004

Moses, J. (2018). *The first modern risk: Workplace accidents and the rise of the social state*. Cambridge University Press.

National Safety Council. (2022). *The most dangerous time to drive.* https://www.nsc.org/road-safety/safety-topics/night-driving

NOAA (U.S. National Oceanographic and Atmospheric Administration). National Weather Service. (2015). *Tsunami awareness and safety.* https://nws.weather.gov/nthmp/documents/nthmpsafety.pdf

Nori, M., & Scoones, I. (2019). Pastoralism, uncertainty and resilience: Global lessons from the margins. *Pastoralism, 9,* 10. https://doi.org/10.1186/s13570-019-0146-8

Norris, F. H., Murphy, A. D., Baker, C. K., & Perilla, J. L. (2004). Postdisaster PTSD over four waves of a panel study of Mexico's 1999 flood. *Journal of Traumatic Stress, 17*(4), 283–292. https://doi.org/10.1023/B:JOTS.0000038476.87634.9b

Nussbaum, M. (2011). *Creating capabilities.* Harvard University Press.

O'Keefe, P., Westgate, K., & Wisner, B. (1976). Taking the 'natural' out of 'natural disaster'. *Nature, 260*(5552), 566–567.

Oxford English Dictionary. (2006). *The Oxford dictionary of phrase and fable* (2nd ed.). Retrieved January 25, 2024, from https://www.oxfordreference.com/view/10.1093/acref/9780198609810.001.0001/acref-9780198609810-e-4825#:~:text=the%20phenomena%20of%20the%20physical,as%20.%20

Parker, L. (1905). *English-Spanish-Pampango dictionary.* American Book and News Co. Publishers.

Prat, S. (2022). Emergence of the genus homo: From concept to taxonomy. *L'Anthropologie, 126*(4), Article 103068.

RAE. (n.d.). *Real Academia Española. Diccionario de la Lengua Española (Spanish dictionary) in Spanish.* Royal Spanish Academy. http://www.rae.es

Redsteer, M., Kelley, K., Francis, H., & Block, D. (2015). Increasing vulnerability of the Navajo people to drought and climate change in the southwestern United States: Accounts from tribal elders. In D. Nakashima, J. Rubis, & I. Krupnik (Eds.), *Indigenous knowledge for climate change assessment and adaptation* (pp. 171–187). Cambridge University Press.

Robeyns, I., & Byskov, M. (2020). The capability approach. In E. N. Zalta & U. Nodelman (Eds.), *Stanford encyclopedia of philosophy* (Summer 2023 ed.). Retrieved January 25, 2024, from https://plato.stanford.edu/archives/sum2023/entries/capability-approach/

Rousseau, J. J. (1967). Rousseau à François-Marie Arouet de Voltaire (Lettre 424, le 18 août 1756). In R. Leigh (Ed.), *Correspondance complète de Jean Jacques Rousseau, Tome IV 1756–1757* (pp. 37–50). Institut et Musée Voltaire, Les Délices. (Original work published 1756)

Sahagún, B. (1979). *Códice florentino. Manuscrito 218–20 de la Colección Palatina de la Biblioteca Medicea Laurenziana (Florentine codex. Manuscript 218–20 from the Palatine collection of the Medicea Laurenziana library)* (3 Vols.). Secretaría de Gobernación, Archivo General de la Nación.

Sapir, E. (1921). *Language: An introduction to the study of speech.* Harcourt, Brace.

Seyedin, H., Habibi Saravi, R., & Hoseini Djenab, V. (2017). Psychological sequels of flood on residents of southeast Caspian region. *Natural Hazards, 88,* 965–975. https://link.springer.com/article/10.1007/s11069-017-2926-z

Shiva, V. (1999). *Biopiracy: The blunder of nature and knowledge*. South End Press.

Shiva, V. (2020). *Reclaiming the commons*. Synergetic Press.

Statista. (2023, November 30). *Life and non-life insurance penetration in selected countries and territories worldwide in 2020 and 2022*. https://www.statista.com/statistics/381174/insurance-penetration-in-selected-countries-worldwide/

USGS (United States Geological Survey). (2022). What is the "triangle of life" and is it legitimate? Frequently asked questions. *Natural Hazards*. Retrieved January 25, 2024, from https://www.usgs.gov/faqs/what-triangle-life-and-it-legitimate

van den Berg, M. (2010). Household income strategies and natural disasters: Dynamic livelihoods in rural Nicaragua. *Ecological Economics, 69*(3), 592–602.

Wallace, R. (2016). *Big farms make big flu*. Monthly Review. Retrieved January 25, 2024, from https://monthlyreview.org/product/big_farms_make_big_flu/

Walters, V., & Gaillard, JC. (2014). Disaster risk at the margins: Homelessness, vulnerability and hazards. *Habitat International, 44*, 211–219. https://www.sciencedirect.com/science/article/abs/pii/S0197397514000824

Whorf, B. (2012). *Language, thought and reality*. MIT Press.

Wisner, B. (1978). *The human ecology of drought Eastern Kenya* [Doctoral dissertation, Clark University, Graduate School of Geography].

Wisner, B. (2016a). Marginality. In P. Bobrowski (Ed.), *Encyclopedia of natural hazards. Encyclopedia of earth sciences series* (pp. 651–652). Springer.

Wisner, B. (2016b, August 31). Vulnerability as concept, model, metric, and tool. *Oxford Research Encyclopedia of Natural Hazard Science*. Retrieved January 25, 2024, from https://oxfordre.com/naturalhazardscience/view/10.1093/acrefore/9780199389407.001.0001/acrefore-9780199389407-e-25

Wisner, B. (2020). Five years beyond Sendai – can we get beyond frameworks? *International Journal of Disaster Risk Science, 11*, 239–249. https://link.springer.com/article/10.1007/s13753-020-00263-0

Wisner, B., Blaikie, P., Cannon, T., & Davis, I. (2004). *At risk: Natural hazards, people's vulnerability and disasters* (2nd ed.). Routledge.

Wisner, B., O'Keefe, P., & Westgate, K. (1977). Global systems and local disasters: The untapped power of people's science. *Disasters, 1*(1), 47–57. https://onlinelibrary.wiley.com/doi/10.1111/j.1467–7717.1977.tb00008.x

Yadav, P., Saville, N., Arjyal, A., Baral, S., Kostkova, P., & Fordham, M. (2021). A feminist vision for transformative change to disaster risk reduction policies and practices. *International Journal of Disaster Risk Reduction, 54*, 102026.

Zachiria, A., & Patel, B. (2006). Deaths related to hurricane Rita and mass evacuation. *The Chest Journal, 130*(4 Suppl. 124S). https://doi.org/10.1378/chest.130.4_MeetingAbstracts.124S-c

Chapter Two
Framing Disaster Risk Reduction

Introduction

For millennia, humans have asked why people suffer: why we grow old, sicken, and die. Hinduism believes suffering results from the consequences (karma) accumulated by behavior in past lives. Buddhism holds that we suffer due to ignorance, grasping, and attachment. In some Christian theologies, it is because we are living in a broken world, a world out of step with divine intentions. Ancient Israelite and Mesopotamian writings suggest a need to justify "innocent suffering" as "an attempt to defend divine justice in the face of aberrant phenomena that appear to indicate a deity's indifference or hostility toward virtuous people" (Bricker, 2001, p. 83). Such exculpatory arguments are known as theodicies. Many Indigenous peoples believe that their ancestors influence their lives, and actions that displease the ancestors will result in suffering. Such meditations on the meaning and ultimate causes of suffering are highly relevant to this chapter but beyond its intention, which is to review some contemporary frameworks that seek to explain only one kind of suffering: loss in the context of impacts from a natural hazard. Indeed, even more narrowly circumscribed, this chapter will review some understandings that draw only on processes, relations, and elements of the material and social worlds in which we live, whilst bracketing all assumed or alleged transcendent realities and assuming the parameters and limitations of our particular universe (Lightman, 2013).

Following on from Chapter 1, "framing" could be defined as the use of language and metaphor to stimulate and sort questions about disaster to reveal those that are most useful for different actors in diverse situations. Some people may inquire "how" and "when" in pursuit of proximal reasons for building collapse, avalanche, death, or livelihood

DOI: 10.4324/9781003292814-2

disruption. Others may want to know "who" and "where" in order to conduct a rescue. Yet other people may want to ask "why" the disaster occurred. The "why" question has had the most varied answers, each framed and guided by the particular interests, motives, and audiences of the questioner. Before continuing in this chapter with an account of the more practical, quotidian, and mundane forms in which the question "why" may appear (and useful answers suggested by different frameworks), we examine more the capital "W" why-questions that will always lurk behind the scenes.

Ultimate but Not Useful Questions

A Buddhist scholar explains (Dhammananda, 2010, online):

> Disaster takes place from time to time because basic elements like water, earth, wind and fire are always in a constant flux and need to find a balance. That is why the Buddha said that a human life is *dukkha* [suffering], because he/she is also subject to such a flux. Regardless of status or species, because of this universal flux, every creature or being experiences *dukkha*.

A Catholic professor of social work says (O'Mathúna, 2018, p. 40):

> The beauty of the world reminds us of God's good side, and disasters remind us of his dangerous side. He created the world with love and beauty, and humans introduced sin and ugliness. This requires a just response, which theology calls God's judgment. . . . The Bible records that this has happened with specific events from time-to-time, but it does not claim that every disaster is the direct act of God.

According to a Nigerian Igbo myth (Kanu, 2014, p. 9. citing Mbaegbu, 2012):

> [T]here was a time the sky lay just above the earth. Thus, human beings could reach *Chukwu* (God) through a rope which hung down from the sky. There was no death at the time God created the world and human beings. Human beings never lacked anything, as God was involved in their affairs: associated with them and provided for all their needs. In that world, the spirits and human beings were one. However, it happened one day that a woman spat into the eyes of God. He picked offence and thus withdrew into the sky where he now dwells. The result is that human beings were punished and could now die. Since then suffering came into the world.

Three Christian evangelicals write (Stiekes et al., 2017):

> [T]he apostle Paul . . . helps to explain the presence of natural disasters. "For we know", says Paul, "that all the creation groans together and suffers the pains of childbirth together until now" (Rom 8:22). . . . Note the universal impact of Paul's claim. "All" creation groans "together" and suffers "together". And the suffering is ongoing. We tend to deliberate about God's purposes in a single event such as a hurricane or an earthquake or a tsunami. But the truth is, the whole world is always groaning in pain every day.

An atheist declares (Dawkins, 1996, p. 132):

> In a universe of electrons and selfish genes, blind physical forces and genetic replication, some people are going to get hurt, other people are going to get lucky, and you won't find any rhyme or reason in it, nor any justice. The universe that we observe has precisely the properties we should expect if there is, at bottom, no design, no purpose, no evil, no good, nothing but pitiless indifference.

Finally, let us hear from Lightman (2013, p. 35), who would agree with Dawkins about the random interaction of fundamental forces in an "accidental" and "disembodied" universe yet who still believes that human sensibility can transform what Dawkins sees as "pitiless indifference":

> Suppose I ask a different question: If against our wishes and hopes, we are stuck with mortality, does mortality grant a beauty and grandeur all its own? Even though we struggle and howl against the brief flash of our lives, might we find something majestic in that brevity? Could there be a preciousness and value to existence stemming from the very fact of its temporary duration?

Seeking to know why people suffer, all of the aforementioned seek answers in transcendent realities that are more-than-natural, even Lightman (2013), for whom the answer is immanent, provided by human desire and sensibility. Only Dawkins sifts through material reality, finding merely "pitiless indifference".

Looking for causes of suffering and loss within the limits of historical time and Cartesian space – that is, within our immediate material and social worlds – can require the unspooling of a complex web of contingent factors involving human behavior and institutions, the use and abuse of power, the built environment and inhabited planet, and numerous geophysical, biological, lithological, hydrological, atmospheric, and astronomical processes, each of them complex in their own right. The frameworks discussed in this chapter all pursue the causes of human suffering in different ways with different goals and audiences in mind. Above all, they all search for useful answers, not the ultimate ones.

Useful but Not Ultimate Questions

The ancient Greek philosopher Aristotle famously taught that the answer to why things move and change in nature will describe one of four causes: final, formal, material, and efficient. Imagine you have pizza and juice you share with your friends. Dining with your mates is the final cause, the telos, or goal. Dining could require a table. The table's material cause might be wood. Its efficient cause is carpentry or the carpenter. The table's design is its formal cause.

In a similar way, one may ask why a disaster occurs by inquiring into its four causes. A consensus emerged in the 1970s and 1980s that disasters are not natural, as discussed in Chapter 1. They always involve social relations and human behavior. In Aristotelian terms, "final cause", or goal, is pursuing happiness or avoiding pain. Every human motive from crass greed to compassionate generosity can be understood as fundamentally one or the other. The "material cause" is the hazard. This is a material thing or process, such as a steep slope or cyclonic storm. The "efficient cause" is the human activity involved in achieving the goal (final cause): draining a wetland, terracing a slope, building a road or a port facility, drilling for gas or oil, or damming a river. The "formal cause" is the design or plan for the execution of the building, terracing, etc., and importantly, this includes the government regulations, safeguards, inspections, and broadly speaking, risk management that arose historically alongside the modern state (Bandopadhyay, 2022).

Of course, Aristotle's causes miss out many intermediate causes and interconnections. The why-question has certainly not been fully answered. Human livelihood activities are complex negotiations with nature and with other human beings. This is true whether livelihood is based on shifting cultivation, camel and goat pastoralism, university lecturing, or wage labor in a pencil factory. We, therefore, need a framework that can generate questions revealing the web of relations with nature and other people and the fluidity and dynamism of that web. One such framework is known as PAR.

PAR: The Pressure and Release/Progression of Vulnerability Framework

PAR distinguishes among processes that are more and less remote in time and space and then traces the cascading consequences of more remote processes as they are focused by more local and situational pressures, eventually ending up influencing immediate safety conditions. This approach began with the work of Ian Davis in the late 1970s that has subsequently been elaborated and applied (Davis, 1978, 1987). This section reviews the development, critiques, refinements, and limitations of the approach he pioneered that was eventually formalized as the pressure and release/progression of vulnerability and resilience (PAR) framework. The discussion will underscore PAR's relevance, usefulness, and usability, while challenging its structuralist and Western bias. This approach, which will be taken up in Chapter 3, is often referred to as decolonizing disaster risk studies and applications.

Origin and Evolution of PAR

PAR can be traced back to Ian Davis' 1978 book, *Shelter after Disaster*. Page three of Davis' 1978 book contains the Ur-diagram that eventually became PAR (Figure 2.1).

Later, Davis (2012) wrote about the origin of this diagram:

> I had been impressed with a paper in *Nature*, "Taking the naturalness out of disasters" (O'Keefe et al., 1976) . . . Considering all this, I began to sketch in an attempt to visualize vulnerability and hazard as a pair of converging arrows, meeting in a disaster, or meeting where there was the potential for a disaster.

As one of the co-authors of *At Risk*, Davis imported his "crunch" diagram into that book (Blaikie et al., 1994) (Figure 2.2).

The revision and expansion of *At Risk* ten years later added detail to the treatment of root causes and dynamic pressures (Wisner et al., 2004). Wisner et al. (2012a) improved PAR by adding the influence of dynamic pressures and unsafe conditions on hazards as a feedback loop in Figure 2.3.

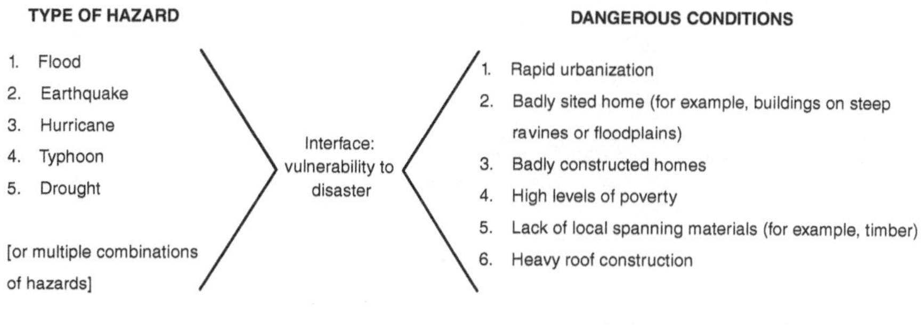

FIGURE 2.1 Hazards and vulnerability

Source: Davis (1978, p. 3)

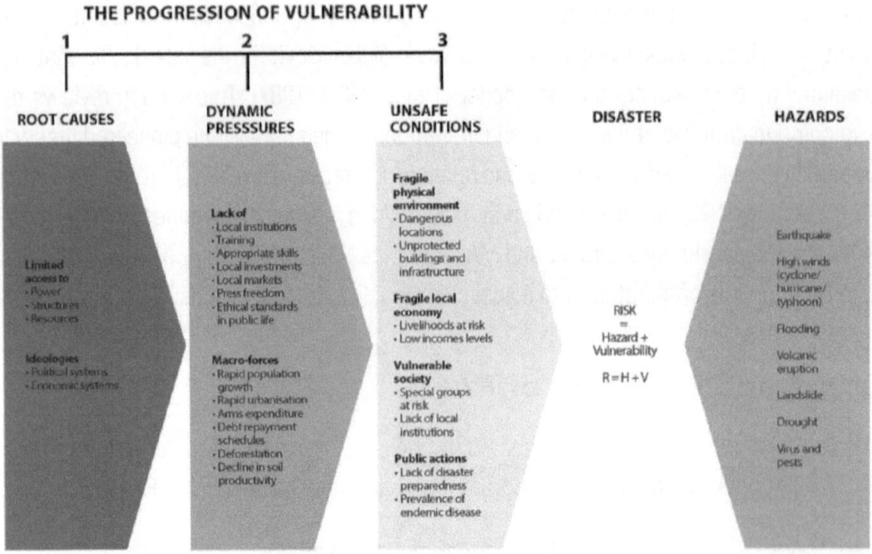

FIGURE 2.2 Progression of vulnerability, 1994 version

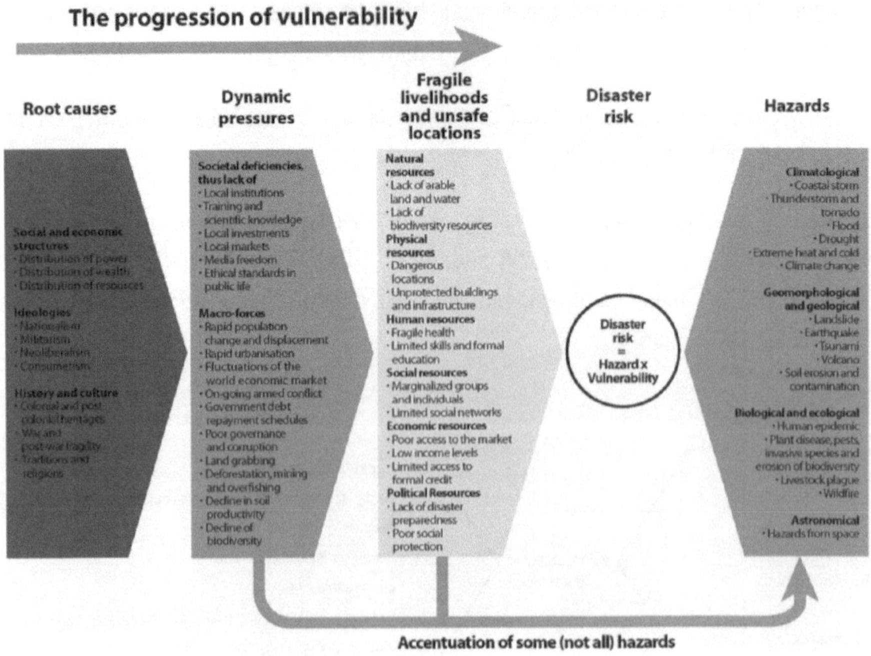

FIGURE 2.3 Progression of vulnerability

Source: Wisner et al. (2012a)

Another improvement by Wisner et al. (2012a, p. 23) was to add "fragile livelihoods" to unsafe locations (conditions). They based this revision of PAR on current work about sustainable livelihoods and, in particular, access to six arbitrary (and, hence, challengeable) kinds of resources – natural, human, social, physical, economic, and political (Wisner et al., 2012a, p. 29). Unfortunately, this revision did not point out the trap lying in mainstream interpretations of "sustainable development" as "sustainable growth" (a contradiction in terms). JC Gaillard further enriched this reconsideration of PAR by contributing two visualizations: the "triangle of vulnerability" (Wisner et al., 2012a, p. 27) updated in Figure 2.4 and the "circle of capacities" (Wisner et al., 2012a, p. 28) updated and presented here as the "circle of abilities" (Figure 2.5). Figures 2.4 and 2.5 perpetuate the Western obsession to categorize, an approach that distorts the reality of the world, as all cultural and linguistic lenses do. Indeed, many of the items listed in each of the same-level boxes in the diagram could stand elsewhere or be worded differently. These challenges are ever-present in communication.

In co-author JC Gaillard's hands, PAR has been interpreted as oppressive and constraining processes in social, economic, and political domains that reduce and limit access to a

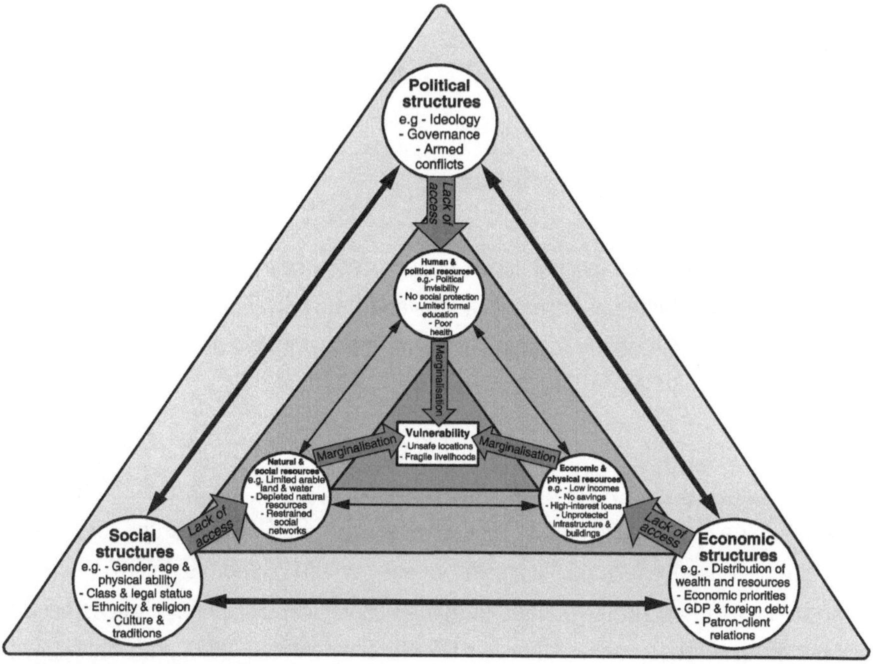

FIGURE 2.4 The triangle of vulnerability

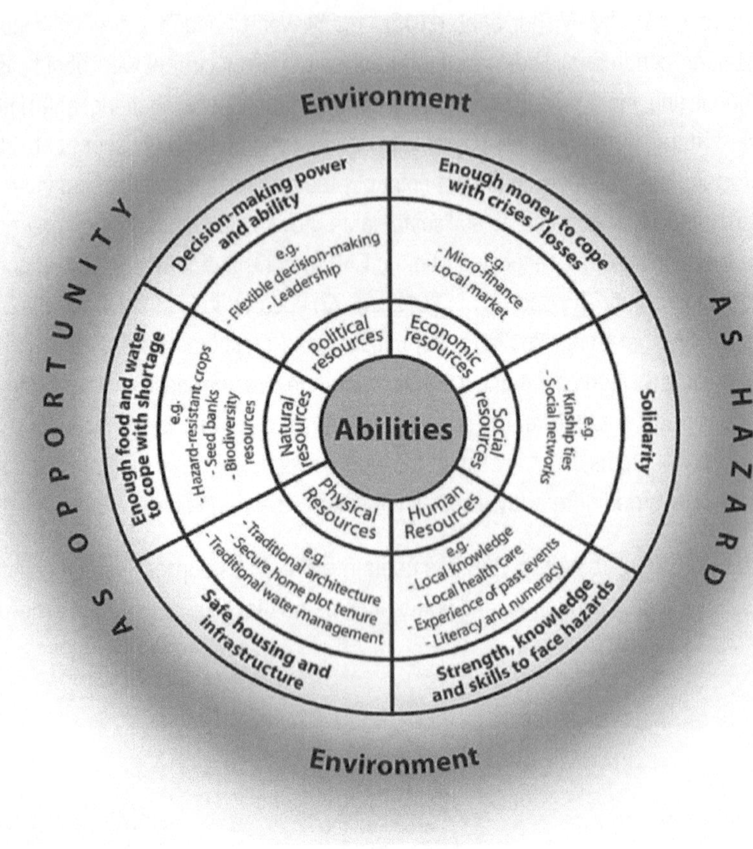

FIGURE 2.5 The circle of abilities

variety of physical and socio-economic and political resources. At the center is the result: vulnerability as manifested in unsafe locations and fragile livelihoods. An advantage of the triangle form is the ability to represent a far more dynamic and interactive reality than the more linear PAR in its earlier renditions.

The other addition, Gaillard's "circle of abilities", is based on a fundamental axiom of Western thought, the binary: the environment as either and both "hazard" and "resource". In Chapter 3, we critique and situate such Eurocentrism. Still, we have our positionality as five authors raised, educated, living, and working within Western thought. This binary sits at the heart of geography and other human ecological disciplines. It goes back to the work of human geographers Barrows (1923), Zimmerman (1933), and White (1945) and then more recently in studies by Burton et al. (1978, 1993). True to that tradition of geography-as-human-ecology, Gaillard means his circle to be read from the inside out, not, as environmental determinists would do, from the outside-in.

Thus, human capacities to cooperate, to learn, and to struggle against obstacles (including more powerful, coercive humans) allow the utilization of resources listed in the inner concentric ring, and these resources, in turn, have the outcomes listed in the outermost ring via skillful means, including those named as examples. So, for example, economic resources are utilized via micro-credit with the result that a person or household has enough money to cope with crises and losses. Note that the circular configuration of the rings correctly implies that mobilization of economic resources requires social, human, physical, natural, and political resources, again, grouped after arbitrary (and, hence, challengeable) categories. Figure 2.5 must be read around and around as well as from the inside out in order to capture the full complexity, interdependence, and dynamism exhibited by human capacity and how capacities allow people to seize the opportunities offered by their living, non-living, and built environments as well as to avoid the hazards these environments present (Chapters 4–8).

Finally, Davis (2015) recently produced a version of PAR that expands inquiry into the role of root (or underlying) causes and dynamic pressures in shaping hazards that unpacks what Wisner et al. (2012a) were hinting at with their feedback arrows. This effort goes a long way to answer one frequent criticism of PAR the reader will encounter in the next section – that PAR neglects the right-hand side of the diagram. Bringing all these updates together, we have produced Figure 2.6.

Criticisms of PAR

Hammer et al. (2019) found that PAR lacked the precision to allow quantification. This shows an unfortunate misunderstanding of what PAR is and what it sets out to achieve. As stated earlier in this chapter, PAR – like all frameworks – seeks to remind users of possibly relevant questions about hazard, vulnerability, and risk. That is its only function. It was never meant to be a quantitative model. Kasperson et al. (2003) and Turner et al. (2003) had earlier expressed similar misgivings with similar misunderstandings, stating that the 1994 version of PAR lacked sufficient concern with the biophysical environment and human interactions with the environment. Indeed, others have found PAR to have paid insufficient attention to exposure as a key component of vulnerability (Hammer et al., 2019), although it could be argued that all aspects of exposure are included in PAR's vulnerability and so another word, exposure, is not needed.

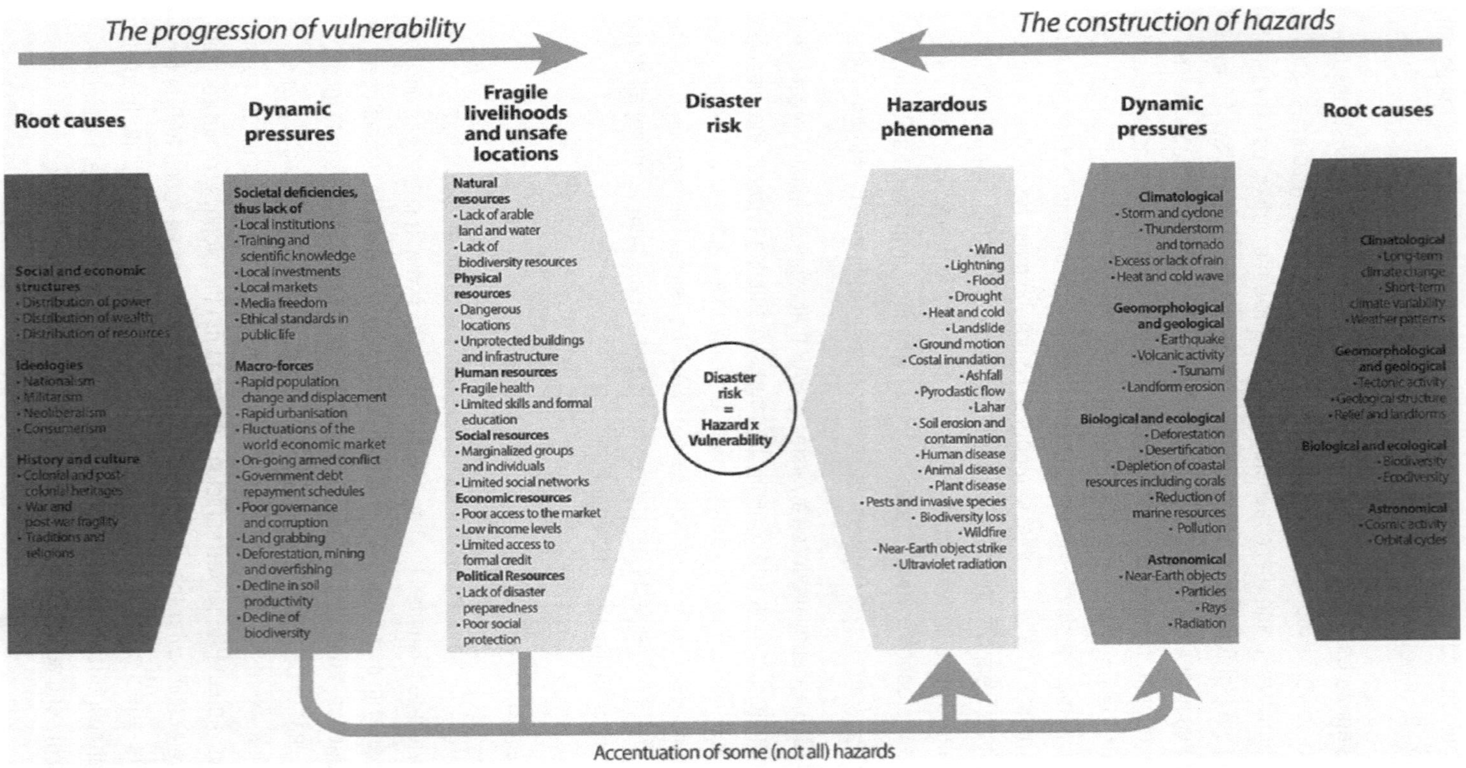

FIGURE 2.6 Progression of vulnerability, 2024 version

Irrespective, the overall concern expressed is PAR's "one-sided" focus on human, social, and political factors displayed earlier on the left-hand side of Davis' diagram. This is a criticism Malm in 2022 voiced in personal communication with Wisner. Malm would like to see a possible "right-hand side" dealing with hazards and their own human social production. This would go further than the symmetry suggested by Davis' (2015) attempt, where the root causes, dynamic pressures, and unsafe conditions he lists on the right-hand (hazard) side are all geophysical or atmospheric processes, not human ones.

Lundgren and Strandh (2022) applied PAR in Mozambique. While echoing some of these criticisms, they found they could mitigate PAR's weaknesses by using it with the access model developed in *At Risk* (Blaikie et al., 1994; Wisner et al., 2004). Lundgren and Strandh write (2022, p. 16):

> The advantage of this model is that it enables a broad understanding of underlying factors creating vulnerability in a specific context. However, the model is to some extent abstract, complicating the distinctions of driving factors. To compensate this limitation [sic] I will attempt to present the temporally and spatially distant findings clearly. Another disadvantage of the model is its lack of explaining why some communities are able to cope with disasters, therefore the analytical framework will include an adaptation of the Access model.

The access model functions at the granular scale of the household, disaggregating hazard impact and human choice and action to that scale. It was first developed and deployed by *At Risk* co-author Piers Blaikie in a study of HIV-AIDS in Uganda (Barnett & Blaikie, 1992) and later introduced *At Risk* (Blaikie et al., 1994), where, in addition to leading a general chapter

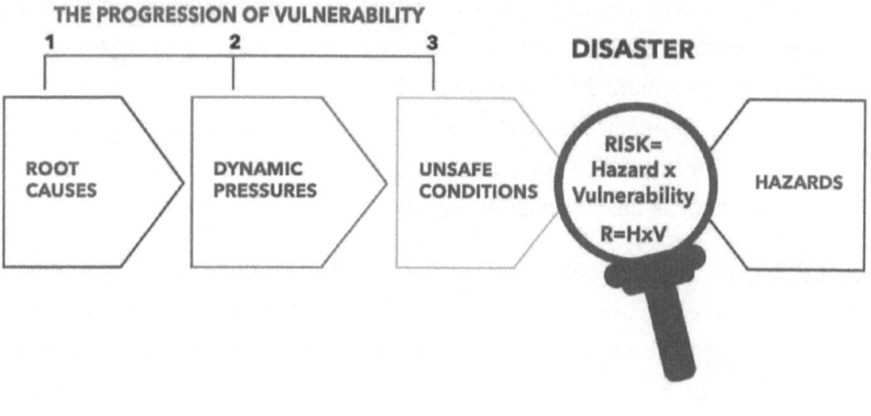

FIGURE 2.7 Progression of vulnerability, 2017 version

Source: Sandoval Henríquez (2017)

on access (Wisner et al., 2004, pp. 49–86), he worked with co-authors to focus another chapter on the role of access in food security and famine (Wisner et al., 2004, pp. 87–124). If one takes PAR with the access model, criticisms melt away that PAR is too abstract, is linear, and lacks sufficient detail to guide interventions and policy. In some versions of the graphical representation of PAR, the stylistic magnifying glass is the location where the dynamic iterations of the household access process take place (Figure 2.7). The conceptual space viewed through the lens of the magnifying glass is the locus of the access model.

Households are the unit of analysis in the access model, explained as numbers (refer to the boxes in Figure 2.8) (Wisner et al., 2004, pp. 88–91):

1. Each household has a livelihood basis and has access to resources and assets (land, labor, capital, tools, networks, and expectations of resources from others). This is the "household access profile". On this basis, households have income opportunities that require access to qualifications and payoffs. All of these processes take place with overarching social relations and structures of domination that are "rules of the game": "who eats first, who will have to absorb consumption cut-backs in times of dearth or who receives medical treatment" (Wisner et al., 2004, p. 102). Household members choose income opportunities that yield a livelihood that becomes part of the household budget. Decisions (including through the social relations and structures of domination) are made about using the budget for consumption, investment, and the realization of assets. There are outcomes to these decisions, which, in addition to any household demographic events (births, deaths, marriages), return one to the beginning of the model.
2. The household and its livelihood profile must negotiate unsafe conditions as a daily reality. Specific hazards (Figure 2.8, item 3) with unique time/space characteristics arise. This may become a "trigger event" (5) that may not only modify social relations and structures of domination (thus affecting social protection) but that also initiate the "transition to disaster" with impacts on "normal" life (6). Households react, cope, adapt, endure, or enjoy other interventions by agents (extended family, neighbors, government) as part of the "disaster process" (7). This leads to the next disaster with or without action for disaster reduction (8).

Fully rebutting the criticism that the approach in both editions of *At Risk* lacks specificity and, therefore, has little or no practical use, this section of the book continues for more than a dozen pages to tease out the links and loops that describe household access (Wisner et al., 2004, pp. 101–115), including a mammoth summary figure, "The Access model in the transition to disaster" (Wisner et al., 2004, pp. 104–105).

Returning to a review of criticisms, Wolbers et al. (2021) fault PAR for being out of sync with many of the issues and topics discussed in what they consider to be "the top ten

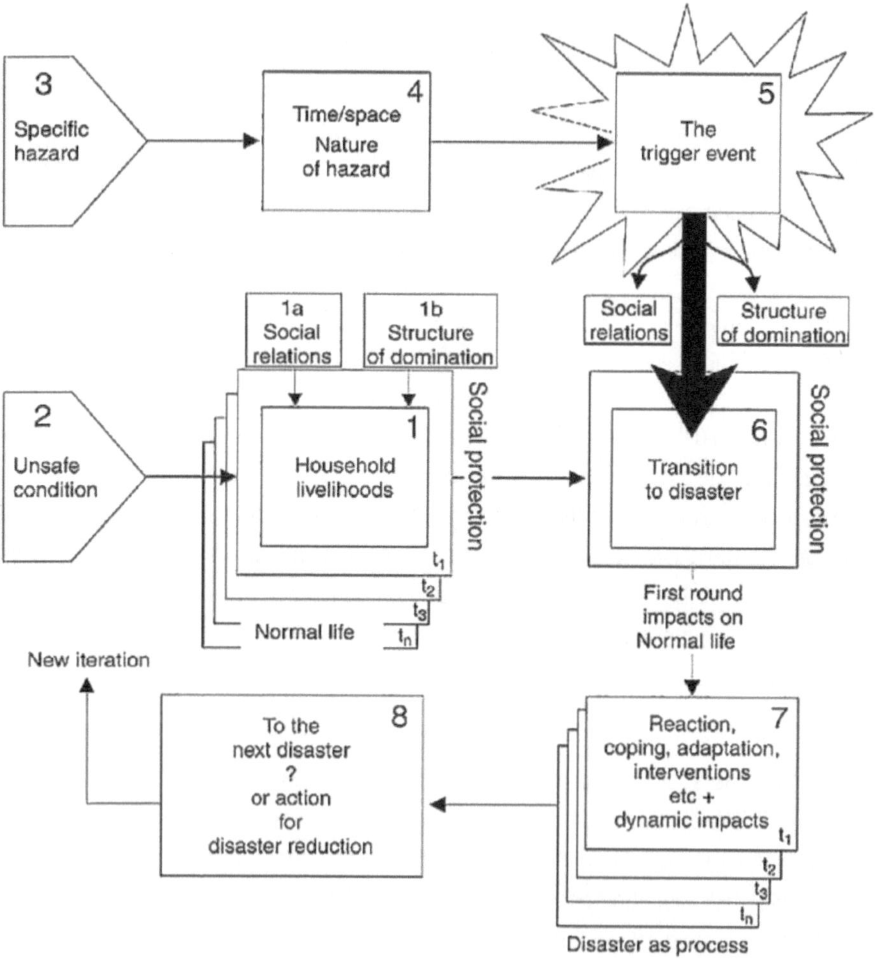

FIGURE 2.8 The access model in outline

Source: Wisner et al. (2004, p. 89)

conceptual papers" in the period 2001–2020. It is certainly the case that *At Risk* is not trendy and does not belong to the outpouring of work specifically focused on climate change (Delina, 2023; IPCC, 2022; Goodel, 2017; Dryzek et al., 2013; Mercer, 2010). Nevertheless, *At Risk* treats climate change at ten places throughout its text and discusses climate change in relation to floods (Wisner et al., 2004, pp. 171–172, 212–213), to biodiversity (Wisner et al., 2004, pp. 195–196) – which, in fact, was far ahead of its time considering recent laments that climate change and biodiversity are insufficiently connected alongside efforts to connect them better – and to hurricanes (Wisner et al., 2004, pp. 246–247).

In addition, concerning more of Wolbers et al.'s "issues and topics", a brief perusal of the chapter titles and index of a book inspired by *At Risk* and co-edited by one of *At Risk*'s authors shows that the forward momentum of work beginning in the 1970s and that led to *At Risk* has not been exhausted (Wisner et al., 2012b). Chapters treat gender, sexuality and disaster; children, youth and disaster; elderly people and disaster; disability and disaster; caste, ethnicity, religious affiliation and disaster; civil society and disaster; community action and disaster; social protection and disaster; economic development policy and disaster; education and disaster; media, communications and disaster; psychosocial recovery; socio-economic recovery; health and disaster; culture, hazard and disaster; and much more.

Another criticism concerns PAR's engagement with gender. Whilst PAR does mention gender, its treatment is superficial in both editions of *At Risk*, and there is no consideration of identity and the situation of sexual and gender minorities. To fill this gap, Oxfam developed a "gendered progression of vulnerability" framework (Hai & Smyth, 2012). Going further, one notes that treating people as risk bearers and agents in the creation of and response to disaster risk neglects people's experience. This is true of approaches focused on the perception of risk that do not tap the subjective experience of living with or being affected by disaster risk. Hewitt (1997) was fast off the mark with this criticism. He is correct. That subjective experience of being "at risk" is not evoked or addressed in either edition of *At Risk*. The PAR framework is too rigidly structural and deterministic to allow either first-person experience or emotion.

Finally, there is the question of PAR's application to concrete situations. Does it actually guide practitioners, or more dramatically stated, does PAR save lives? Addressing this question, Michellier et al. (2020, p. 317) found that PAR showed "discrepancies between [the] conceptual model[s] and the basic structures of international programs that deal with disaster risk reduction and fieldwork observations". In other words, PAR may be difficult or impossible for guiding actual practice. In another critic's words, "there is not sufficient detail to provide adequate structure and explanation of system behaviour" and that PAR is linear, therefore, "too simple to be useful in risk and vulnerability analysis" (Wilder, 2018, p. 51).

Fair enough; however, here again, the criticism misses the point of PAR. PAR is meant to nudge or even goad researchers, policy makers, and practitioners to consider key questions about root causes, dynamic pressures, unsafe conditions, and hazards. It does not suggest specific actions to be taken once people have responded to the challenge of PAR and asked such key questions. Moreover, PAR has a much-neglected fraternal twin. Remember that the abbreviation "PAR" is short for "pressure and release", or the "progression of vulnerability". In Chapter 10 here, its sibling appears, a framework dedicated to the progression of safety, sometimes called RELEASE for "release of pressure". RELEASE serves the same mnemonic function as PAR: to stimulate questions. As will be seen in Chapter 10, the answers to key questions lead to precisely the locally appropriate policies and practices that critics assert are missing in an overly abstract PAR.

Applications of PAR

Wisner and Alcántara-Ayala (2023) reported on the applications of PAR using the search engine Scopus. Using the name of the publication in which the PAR was first released, the first and second editions of the book (Blaikie et al., 1994; Wisner et al., 2004) during the period 1994–2021 were considered. As of August 2022, a total of 5,423 publications were found. The software WordStat 9 (Provalis, 2023) was used to conduct a quantitative content analysis of the titles and abstracts of the selected papers. Although most publications that refer to the PAR framework are in English (96%), it has been referenced in 15 other languages.

As shown in Figure 2.9, the number of the references made to the PAR framework increases over time. From 2011–2012, the number of references increased by 170%. Still, it is hard to draw definitive conclusions without normalizing the data by difficult-to-obtain denominators, with examples being the number of researchers and publications in a specific field. Some might suggest that high-profile disasters increase engagement with PAR and disaster research more widely, such as 2005's increase being (partly) attributable to the 26 December 2004 tsunami and 2012's increase being (partly) attributable to both the 12 January 2010 Haiti earthquake and the 11 March 2011 Japan tsunami. Conclusions about any trends in absolute numbers must be viewed cautiously and must be interpreted without the context of normalizing the absolute numbers in different ways.

The first identified reference was published the same year the PAR framework was released. It focused on analyzing the concept of vulnerability, which was applied to a case study of refugee vulnerability in Greece. The study recognized that, in addition to the characteristics of individuals and households, consideration must be given to the context of state policies and the wider characteristics of society and the economy to understand the dimensions of vulnerability (Black, 1994).

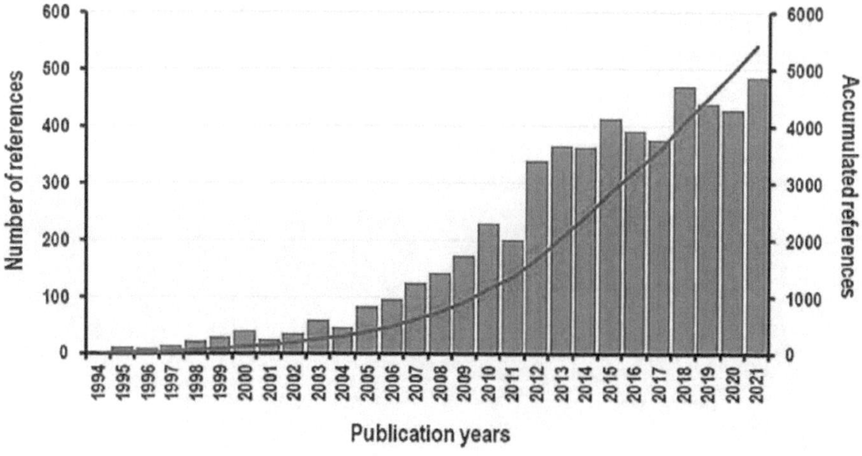

FIGURE 2.9 Temporal distribution of the references made to the PAR model during 1994–2021

During the following year, in 1995, a series of papers considered the PAR framework. Chan (1995) investigated the effectiveness of government-run permanent relocation schemes in response to flood hazards in Malaysia. Chiotti and Johnston (1995) were already concerned about global environmental change in the context of farm adaptation. Based on a literature review, they suggested that a better understanding of the nexus between climate change and farm adaptation requires an approach that situates farm-level decision-making within structural forces, including biophysical ones. Also in 1995, eight out of thirteen papers in a special volume of *GeoJournal* entitled "Questioning Development: Growth? – Destruction? – Sustainability?" (Vol. 35, No. 2, February 1995) included insights from the PAR framework (Yapa & Wisner, 1995).

Over the years, the core research themes for which the PAR framework has been used have evolved. Climate change, disaster risk, social vulnerability, disaster risk reduction, natural hazards, and risk management were at the top of the most frequent keywords included in the abstracts of the publications reviewed (Figure 2.10). As always, this visualization must be viewed with caution due to the lack of normalizing calculations and its de-contextualization from other circumstances of academic publication.

The journals in which the largest number of PAR-relevant research was published were the *International Journal of Disaster Risk Reduction*, *Natural Hazards*, and *Disasters* (Figure 2.11) All of them are indexed in the Web of Science database, which only includes peer-reviewed publications. This graph does not factor in the number or length of papers published by each journal – *International Journal of Disaster Risk Reduction* in particular has been publishing large numbers of manuscripts – or the journals' mandates since some are more empirical or interdisciplinary than others.

Examining the publications by geographical location of the author shows the concentration of references to the PAR framework in 21 countries, in each of which at least 150 articles were published (Figure 2.12). Although this illustration does not in any way reflect the countries where the PAR framework is most valued, these figures indicate

FIGURE 2.10 Visualization of the most frequent words used in abstracts of the publications referring to the PAR framework

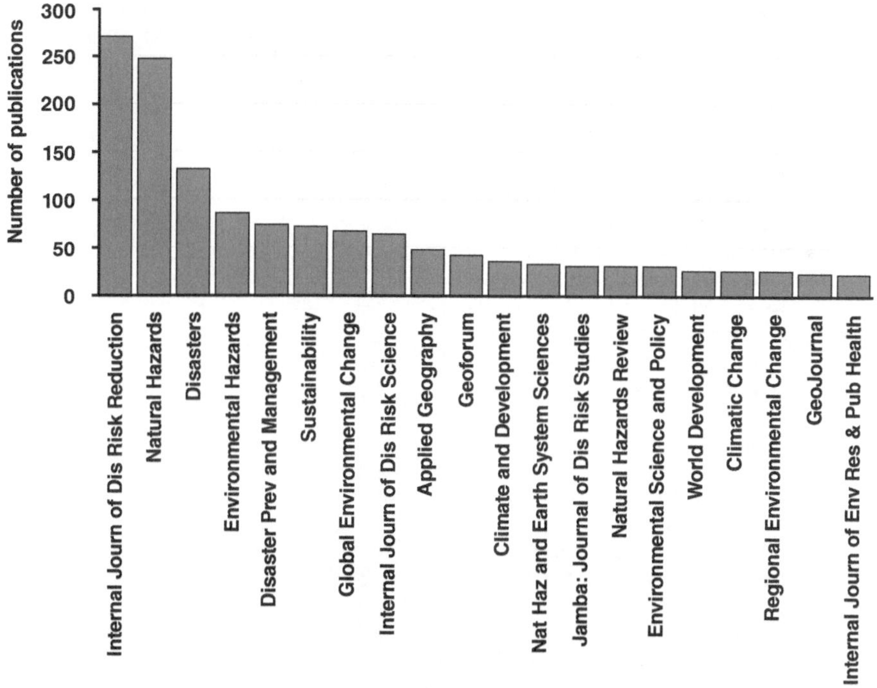

FIGURE 2.11 Frequency of publications referring to PAR by journal of publication

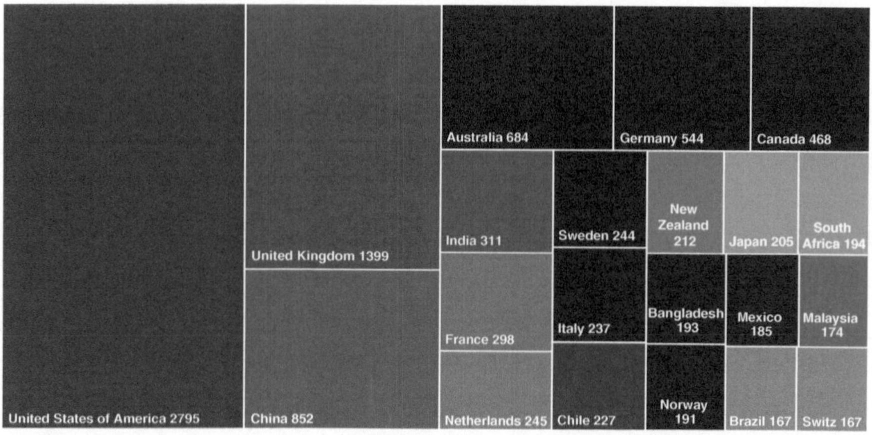

FIGURE 2.12 Frequency of publications referring to PAR by country of affiliation of the author

the un-normalized geographical distribution of the research carried out and published. Figure 2.12 reveals a few large regional scientific hubs in the United States of America, the United Kingdom, China, and Australia.

It was also possible to identify a handful of academic institutions where the staff produced the largest number of publications referring to the PAR framework (Figure 2.13):

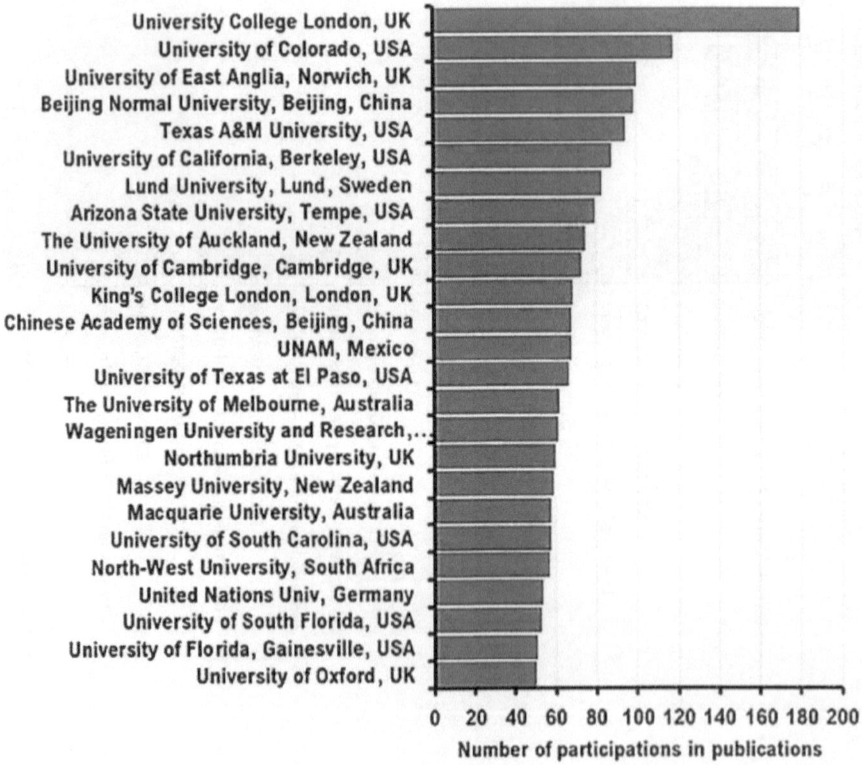

FIGURE 2.13 Number of authors in publications referring to the PAR framework by author institution

University College London (UK), University of Colorado (USA), University of East Anglia (UK), Beijing Normal University (China), and Texas A&M University (USA). Figure 2.13 does not account for universities having the resources to pay publication charges or other factors influencing the ability to publish in peer-reviewed journals, irrespective of research interests and the research conducted.

Other Frameworks and Approaches

Capacity and Vulnerability Focused Approaches – Oxfam and La Red – 1987–1996

Oxfam has been an incubator of innovative work on disaster for many decades. A foundational way of looking at vulnerability and capacity began as a report for this NGO (Anderson & Woodrow, 1987). According to this view, variations in vulnerability to natural hazards could be balanced and counteracted by capacities. Then Oxfam published

Maskrey's (1989) study based on work in Peru. This slim volume has had an outsized influence. Maskrey (1989) similarly adopted a capacity and vulnerability (CVA) framework, as did a number of the people he worked with in Latin America and the Caribbean (LAC), where there were many researchers and organizations working on disaster risk.

In 1992, several individuals and organizations in the LAC region created The Network of Social Investigation for the Prevention of Disasters (*La Red de Estudios Sociales en Prevención de Desastres*) or *La Red* for short (La Red, 1992). The framing of disaster risk developed at this time in LAC accepted that disasters were in no way natural. They were "socially constructed" (Maskrey, 1993). Furthermore, disaster was "a failure of development", a distortion of what they assumed was a normal historical trajectory. Development and disaster risk reduction were identical, and being so, *La Red* welcomed the mainstreaming of "disaster risk reduction" (DRR) by international organizations from UNDRO in the late 1970s through to the work of the UN Office for Disaster Risk Reduction (UNDRR) (UNDRR, 2022; Haghebaert & Kelman, 2024).

The members and associates of *La Red* not only developed and diffused new thinking about disaster risk but also conducted training courses, advocating and lobbying for the application of new thinking in government policy and practice. Over the years, the thinking of various founders of *La Red*, such as Gustavo Wilches-Chaux in Colombia, Allan Lavell in Costa Rica, and Virginia García-Acosta (1996, 1997, 2008) and Elisabeth Mansilla (1996) in Mexico, exerted influence among government officials and on regional organizations, such as CEPREDENAC. The latter is the specialized institution of the Central American Integration System for Disaster Prevention, Mitigation, and Response, founded in 1998. In 2015, one founder, Allan Lavell, won the Sasakawa Prize awarded by UNDRR for contributions to disaster risk reduction.

La Red's influence extended beyond the region, and the writing of *At Risk* was nourished by contact with this group. In fact, a Spanish translation of *At Risk* was facilitated by *La Red* and appeared in 1996 (Blaikie et al., 1996) (Figure 2.14). *La Red* further translated key Anglophone texts from the North American literature (Lavell, 1994).

There was an appreciation of the mutual influences that the work of European and Latin American disasterologists/*desastralogos* manifested. The Spanish preface to the translation of *At Risk* states (Maskrey, 1996, p. 8):

FIGURE 2.14 Cover of *At Risk*'s Spanish translation

[T]he translation and publication of such an important book as Vulnerability: the Social, Political and Economic Environment of Disasters was of utmost importance for La Red. We believe that it is essential that the conceptual approach raised by the authors be disseminated and discussed widely in Latin

> America. . . . [O]ur goal is to ensure that endogenous research in the region is not carried out in isolation without being placed in the context of other currents of thought. We believe that a global criticism such as the one presented in *At Risk* cannot be absent from desktops and libraries of disaster researchers in Latin America.
>
> Likewise, La Red shares with the authors of *At Risk* a concern that research influences the development of more appropriate local, national and international management policies. The first years of the UN's International Decade for Disaster Risk Reduction (IDNDR) was dominated by the physicalist conception of disasters that equated disasters with the hazards that trigger them. However, in 1994 the World Conference on Environmental Reduction of Natural Disaster assumed a more social approach to the problem. Despite this shift, it is evident that the policies and programs implemented by most national organizations and international bodies responsible for disaster management continue to rely on physicalist conceptions and pay little attention to the social processes that generate and accumulate vulnerabilities.
>
> Therefore, it is extremely important that *La Red* and others seek negotiation mechanisms that allow the research results to be transformed into policy instruments. We believe that the appearance on stage of *At Risk* in Spanish represents a challenge and provides a tool to improve prevailing vulnerability management policies.

The last paragraph underscores an important difference between members of *La Red* plus its wider admirers/supporters and mainstream academic disaster researchers. The former were passionate about reform, beginning with local government, and worked hard at advocacy, as well as conducting training courses, publishing guidelines, creating materials for schools, and making videos.

The Space of Vulnerability – Circa 1993

Watts and Bohle (1993) provided an influential, thoughtful account of vulnerability in the context of food security and famine (Watts & Bohle, 1993). This framework inspired Gaillard's "triangle of vulnerability" (Figure 2.4) and "circle of abilities" (Figure 2.5). The reasons is that it covers the same political and economic relations and contains an echo of social relations because "entitlement" may extend to access to endowments (food, in the cases that are Watts and Bohle's (1993) focus) allocated within kinship and extended family networks.

"Re-framing" Disaster Risk – Circa 2001

Christoplos et al. (2001) made an early attempt at framing, or rather, re-framing, disaster risk and disaster risk mitigation. They argued that disaster risk mitigation was "falling between cracks" because "disaster mitigation and preparedness have neither the allure of directly 'saving lives', nor of providing an 'escape from poverty' ". They welcomed what they saw as signs that risk management was becoming a mainstream concern. They point to humanitarian organizations recognizing "factors that do not fit into traditional slots on the relief-development continuum, the rising economic costs of disasters" as well as "a growing acknowledgement that aid will never cover more than a small fraction of the costs of disasters". They also point out a rising consciousness among development and humanitarian workers that dealing with risk and insecurity is central to how poor people develop their livelihood strategies.

A Framework for Vulnerability Analysis in Sustainability Science – Circa 2003

This framework has had high visibility because it was published in the journal *Proceedings of the National Academy of Sciences of the United States of America*. Despite its pedigree, close scrutiny raises some questions about its assumptions and usefulness. Its boxes and arrows suggest a systems-analytic approach that aspires to be comprehensive. However, it is not at all comprehensive. The user is distracted from absent power relations that, in reality, permeate the social world. Power in its political, economic, administrative, and coercive forms are invisible. The framework fails to give visibility to subjective experience. A nod given to "global political economy" as a "human influence outside the place" in the visualization of this framework is the only possible reference to power to be found (Figure 2.15). The term "science" in the 2003 publication's title likely refers to what we call Western science, which has been discussed in Chapter 1 and will be critiqued again in Chapter 3.

Accounting for the "sensitivity" and "resilience" (both terms narrowly defined) of the system exposed to hazards requires "revisions and enlargements in the basic design of vulnerability assessments". This, in turn, has practical and operational consequences for those who "treat" (e.g., manage) so-called "coupled human-environment systems" (i.e., reality), such as governments, legislators, and regulators. Such actors are not spelled out in the text.

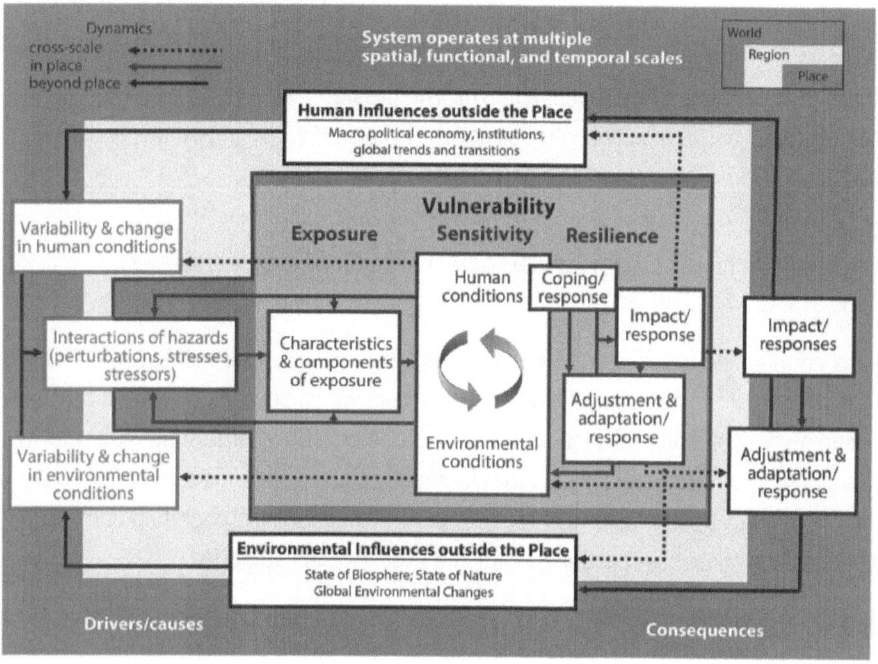

FIGURE 2.15 Vulnerability framework. Components of vulnerability identified and linked to factors beyond the system of study and operating at various scales

Source: Turner et al. (2003, p. 8076)

Alternatives for Sustained Disaster Reduction – Circa 2010

One alternative is an ambitious trilingual polemic and programmatic statement by 20 researchers (including three of us) from around the world (Gaillard et al., 2010). The English and Spanish portions summarize the longer French part. The polemic focuses on critiquing a discourse of extremes ("paradigm of the extreme", Gaillard et al., 2010, p. 67) that privileges hazards as "natural" and encourages a militaristic and colonial approach to risk reduction. Emphasis on extreme events misdirects research and policy attention away from the small and medium disasters, whose total impact on lives and livelihoods can be, in fact, greater. Mainstream practitioners, therefore, miss the plight of many people who have been marginalized by such discourse.

The group of 20 dig into detail, specifying that "the poor, many women, children, elderly, those without shelter, handicapped, refugees, prisoners or members of ethnic minorities are forgotten in the analysis of disaster" (Gaillard et al., 2010, p. 68). Influenced by PAR

(Wisner et al., 2004), marginalization is linked in this analysis to issues of access to resources and the means of production (p. 69) as well as "erosion of entitlement and access to the means of protection" (p. 70). The programmatic part of this document is short and emphasizes locally-controlled actions in concert with sympathetic and accountable local government support (p. 71).

Endangerment Impoverishment Framework – Circa 2012

Lewis and Kelman (2012) cite PAR as well as Hewitt (1983, 1997) and Lewis (1999) as their entry point for laying out a typology of ways in which people's vulnerability may be caused by actions and activities perpetrated by others. They write that "behavior by a few politically or commercially powerful frequently causes endangerment to or impoverishment of many others, as is illustrated by seven categories incorporating bad and ugly behaviors and consequent actions". They then specify the categories:

ENDANGERMENT

1. Environmental degradation.
2. Discrimination.
3. Displacement.

IMPOVERISHMENT

4. Self-seeking public expenditure.
5. Denial of access to resources.
6. Corruption.
7. Siphoning of public money.

This is followed by detailed examples from many countries.

Framing Vulnerability, Risk, and Societal Responses: The MOVE Framework – Circa 2013

This document was a programmatic proposal (Birkmann et al., 2013). The outcome the authors desire is more awareness by policy makers of the links between risks from climate change and other phenomena, as well as the measures available to reduce those risks. The authors promote their framework as "a tool for communicating complexity and stresses the need for societal change in order to reduce risk and to promote adaptation"

(Birkmann et al., 2013, p. 194). Birkmann et al. (2013) do not take the polemical stance of "Alternatives" (Gaillard et al., 2010) but do recognize "the need for social change" in order to facilitate risk reduction and adaptation to climate change. The "tool" they offer purports to integrate "new understanding of coupling, adaptation and resilience" (Birkmann et al., 2013, p. 197), and it would build "a potential ground for closer linking between the concepts and assessment methodologies being developed in DRR [disaster risk reduction] and CCA [climate change adaptation] communities". By not accounting for key prior DRR work and perpetuating the artificial division between DRR and CCA, MOVE offers little success regarding its objectives.

Forensic Analysis of Disasters (FORIN) – Circa 2016

Oliver-Smith et al. (2016) narrate many applications of their 2016 guidelines for the forensic analysis of disasters (FORIN). The term "forensic" refers to a systematic and analytical investigation to seek root causes and drivers of disasters and disaster risk. They argue that within the mosaic of skewed, class-biased, and class-based development practices that continue to affect all societies, there is a need to address the meandering interplay that shapes the social construction of risk. By the same token, the FORIN approach enables researchers and practitioners to explore alternative opportunities for addressing interactions and interdependencies among disaster risk drivers, their systemic nature and challenging complexity (Figure 2.16). Although not differing from what (Western) scientific research ought to be and often has been over past centuries, the FORIN perspective involves an overall strategy to possibly stimulate the policy making domain to understand causes and effects. Much such work is still being done in that vein (Alcántara-Ayala et al., 2023; Ferreira et al.,2023). For instance, Fraser headed a team that applied FORIN to a number of flood-prone locations (Fraser et al., 2016).

Preparing for Extreme and Rare Events in Coastal Regions (PEARL) – Circa 2020

The PEARL framework developed by Fraser et al. (2016) during a large flood risk project provides an additional perspective on root causes of disaster risk (https://cordis.europa.eu/project/id/603663/reporting/de). Fraser et al. (2016) broaden FORIN's insistence on the relevance of governance processes when addressing other underlying factors (Fraser et al., 2016). She and her team state that disaster "events" are not discrete and unique; rather, they concatenate and cascade, influencing risk accumulation. PEARL clarifies that the structure of vulnerabilities and disaster risk are strongly interlaced with politics

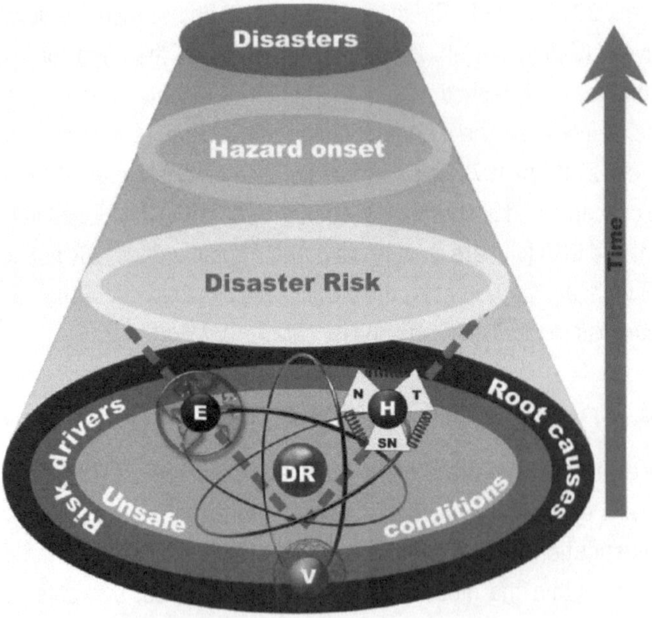

FIGURE 2.16 From social construction of risk to social production of disaster
Source: Oliver-Smith et al. (2016, p. 31)

and cannot be divorced from policy making that concerns multi-scale and multi-sectoral stakeholders and the influence of informal institutional practices. Therefore, Fraser et al. (2020) promote working from disaster risk management "out" and development actors "in". Both reflections on FORIN and PEARL cite PAR.

Discussion and Conclusions

PAR is not the only framework that attempts to understand the factors that create and distribute risk. Still, it has been used often and widely by workers in many fields around the world. A preliminary and partial review found more than 5,000 books and other publications that refer to PAR and 13,000 citations of the publication where PAR was first described in complete detail (Blaikie et al., 1994). One finds applications of PAR at many scales and focused on a variety of problems. These include applications of the following varieties:

- Community-based participatory research.
- Decentralized institutional action, such as local government planning.

- National scale preparedness.
- International development and disaster recovery work.
- Sectoral interventions benefitting health care, school safety, water supply, transportation, and others.

Some criticisms of PAR have been answered. Others remain apposite. Taken separately from the complementary access model, PAR is indeed structuralist, linear, and somehow rigid despite feedback loops bolted on in recent version (e.g., Wisner et al., 2012a). Of course, one must consider why the creators and elaborators of PAR and its extensions, and the authors and readers of *At Risk*, are drawn to these topics, problems, and puzzles in the first place. Few might use words like "empathy" and "compassion" to describe what draws them to work on disaster risk. Since we are all embodied living beings who share that experience and consciousness of our vulnerability to injury, loss, and eventual death, it may be impossible to step aside for theorizing and studying disaster risk with complete objectivity. We and our loved ones are affected. We care. What the critics who say that PAR lacks the subjective element may be missing is the implicit acceptance by PAR's creators of their own vulnerability and the fact that they care.

PAR was never meant as a functional model of social and biophysical reality but as a framework that organizes the world in a way meant to stimulate and encourage specific sorts of questions (Wisner, 2016). Metaphors also function in this manner. We will, therefore, conclude with an extended metaphor with apologies for readers allergic to them or, worse, averse to puns. There exist, in fact, many other frameworks that focus on hazards, vulnerability, and risk. Hopefully, they will all bloom and blossom in their season whilst pollinators move among them and cross-fertilize.

Flawed and incomplete as it is, PAR was only ever meant as one among other frameworks that goad, stimulate, and provoke questions and try to invite criticism and reaction. That is the way knowledge grows, and eventually, a sprout of wisdom appears among the weeds. Under the weight of those criticisms, the creators of PAR should be content to see it settle into the compost that nourishes the conceptual garden. We gardeners could then chant, "PAR-R.I.P." – let our framework rest in peace. Decomposing into the compost and mulch would fertilize the growth of a large farm filled with PAR-SNIPs – activist sociolinguistic "tubers" that "Spread and Nourish Intellectual Property", without boundary fences (to keep out the copyrighting rabbits) (Figure 2.17).

FIGURE 2.17 Concluding parsnips

References Cited

Alcántara-Ayala, I., Burton, I., Lavell, A., Oliver-Smith, A., Brenes, A., & Dickinson, T. (2023). Forensic investigations of disasters: Past achievements and new directions. *Jàmbá: Journal of Disaster Risk Studies, 15*(1). https://doi.org/10.4102/jamba.v15i1.1490

Anderson, M., & Woodrow, P. (1987). *Rising from the ashes: Development planning in times of disaster.* Oxford University Press/Oxfam America.

Bandopadhyay, S. (2022). *All is well: Catastrophe and the making of the normal state.* Oxford University Press.

Barnett, T., & Blaikie, P. (1992). *AIDS in Africa: Its present and future impact.* Guilford Press.

Barrows, H. H. (1923). Geography as human ecology. *Annals of the Association of American Geographers, 13*(1), 1–14.

Birkmann, J., Cardona, O. D., Carreño, M. L., Barbat, A. H., Pelling, M., Schneiderbauer, S., Kienberger, S., Keiler, M., Alexander, D., Zeil, P., & Welle, T. (2013). Framing vulnerability, risk and societal response: The MOVE framework. *Natural Hazards, 67*(2), 193–211. https://doi.org/10.1007/s11069-013-0558-5

Black, R. (1994). Livelihoods under stress: A case study of refugee vulnerability in Greece. *Journal of Refugee Studies, 7*(4), 360–377. https://doi.org/10.1093/jrs/7.4.360

Blaikie, P., Cannon, T., Davis, I., & Wisner, B. (1994). *At risk: Natural hazards, people's vulnerability and disasters*. Routledge.

Blaikie, P., Cannon, T., Davis, I., & Wisner, B. (1996). *El Entorno Social, Político y Económico de los Desastres*. La Red. https://www.desenredando.org/public/libros/1996/vesped/

Bricker, D. P. (2001). Innocent suffering in Egypt. *Tyndale Bulletin, 52*(1), 83–100.

Burton, I., Kates, R., & White, G. (1978). *The environment as hazard*. Oxford University Press.

Burton, I., Kates, R., & White, G. (1993). *The environment as hazard* (2nd ed.). Guilford.

Chan, N. (1995). Flood disaster management in Malaysia: An evaluation of the effectiveness of government resettlement schemes. *Disaster Prevention and Management, 4*, 22–29. https://doi.org/10.1108/09653569510093405

Chiotti, Q., & Johnston, T. (1995). Extending the boundaries of climate change research: A discussion on agriculture. *Journal of Rural Studies, 11*(3), 335–350. https://doi.org/10.1016/0743-0167(95)00023-G

Christoplos, I., Mitchell, J., & Liljelund, A. (2001). Re-framing risk: The changing context of disaster mitigation and preparedness. *Disasters, 25*(3), 185–198. https://doi.org/10.1111/1467–7717.00171

Davis, I. (1978). *Shelter after disaster*. Oxford Polytechnic Press.

Davis, I. (1987). Safe shelter within unsafe cities: Disaster vulnerability and rapid urbanization. *Open House International, 12*(3), 5–15. https://pesquisa.bvsalud.org/portal/resource/pt/des-629

Davis, I. (2012). Box 3.1: Origin of the pressure and release framework. In B. Wisner, JC. Gaillard, & I. Kelman (Eds.), *The Routledge handbook of hazards and disaster risk reduction* (p. 26). Routledge.

Davis, I. (2015). The vulnerability challenge. In I. Davis (Ed.), *Disaster risk management in Asia and the Pacific* (pp. 65–108). Routledge.

Dawkins, R. (1996). *River out of Eden: A Darwinian view of life*. Basic Books.

Delina, L. (2023). *COVID and climate emergencies in the majority world*. Cambridge University Press.

Dhammananda, K. (2010, December 26). The Buddhist explanation of natural disasters, Dharma Records. https://dharma-records.buddhasasana.net/texts/the-buddhist-explanation-of-natural-disasters

Dryzek, J., Norgaard, R., & Schlosberg, D. (Eds.). (2013). *Climate change and society*. Oxford University Press.

Ferreira, A. M., Marchezini, V., Mendes, T. S. G., Trejo-Rangel, M. A., & Iwama, A. Y. (2023). A systematic review of forensic approaches to disasters: Gaps and challenges. *International Journal of Disaster Risk Science, 14*, 722–735. https://doi.org/10.1007/s13753-023-00515-9

Fraser, A., Paterson, S., & Pelling, M. (2016). Developing frameworks to understand disaster causation: From forensic disaster investigation to risk root cause analysis. *Journal of Extreme Events, 3,* 2. https://doi.org/10.1142/S2345737616500081

Fraser, A., Pelling, M., Scolobig, A., & Mavrogenis, S. (2020). Relating root causes to local risk conditions: A comparative study of the institutional pathways to small-scale disasters in three urban flood contexts. *Global Environmental Change, 63,* 102102. https://doi.org/10.1016/j.gloenvcha.2020.102102

Gaillard, JC., Wisner, B., Benouar, D., Cannon, T., Créton-Cazanave, L., Dekens, J., Fordham, M., Gilbert, C., Hewitt, K., Kelman, I., Morin, J., N'Diaye, A., O'Keefe, P., Oliver-Smith, A., Quesada, C., Revet, S., Sudmeier-Rieux, K., Texier, P., & Vallette, C. (2010). Alternatives pour une réduction durable des risques de catastrophe. *Human Geography, 3*(1), 66–88.

García-Acosta, V. (Ed.). (1996). *Historia y desastres en America Latina I.* Centro de Investigaciones y Estudios Superiores en Antropología Social / La Red. https://www.desenredando.org/public/libros/1996/hydv1/HistoriaYDesastresVol_I-1.0.0.pdf

García-Acosta, V. (Ed.). (1997). *Historia y desastres en América Latina II.* Centro de Investigaciones y Estudios Superiores en Antropología Social / La Red. https://www.desenredando.org/public/libros/1997/hydv2/hydv2-todo_sep-09-2002.pdf

García-Acosta, V. (Ed.). (2008). *Historia y desastres en América Latina III.* Centro de Investigaciones y Estudios Superiores en Antropología Social / La Red. https://www.desenredando.org/public/libros/2008/hyd/Historia_y_Desastres_VolumenIII.pdf

Goodel, J. (2017). *The water will come: Rising seas, sinking cities, and the remaking of the civilized world.* Little and Brown Company.

Haghebaert, B., & Kelman, I. (Eds.). (2024). Disaster risk reduction pioneer interviews – a special issue. *Disaster Prevention and Management, 33*(1), 1–2.

Hai, V. M., & Smyth, I. (2012). *The disaster CRUNCH model: Guidelines for a gendered approach.* Oxfam. https://oxfamilibrary.openrepository.com/bitstream/handle/10546/247511/the-disaster-crunch-model-010512-en.pdf?sequence=4&isAllowed=y

Hammer, C., Brainard, J., Innes, A., & Hunter, P. R. (2019). Re-conceptualising vulnerability as a part of risk in global health emergency response: Updating the pressure and release model for global health emergencies. *Emerging Themes Epidemiology, 16*(2). https://ete-online.biomedcentral.com/articles/10.1186/s12982-019-0084-3

Hewitt, K. (1983). The idea of calamity in a technocratic age. In K. Hewitt (Ed.), *Interpretations of calamity: From the viewpoint of human ecology* (pp. 3–32). Allen & Unwin Inc.

Hewitt, K. (1997). *Regions of risk: A geographical introduction to disasters.* Routledge.

IPCC (Intergovernmental Panel on Climate Change). (2022). *Climate change 2022: Impacts, vulnerability, adaptation.* Cambridge University Press.

Kanu, I. A. (2014). Suffering in Igbo-African ontology. *Research on Humanities and Social Sciences, 4*(5), 8–13. https://www.iiste.org/Journals/index.php/RHSS/article/view/11427/11766

Kasperson, J., Kasperson, R., Pidgeon, N., & Slovic, P. (2003). The social amplification of risk: Assessing fifteen years of research and theory. In N. Pidgeon, R. Kasperson, & P. Slovic (Eds.), *The social amplification of risk* (pp. 13–46). Cambridge University Press.

La Red. (1992). *Agenda de Investigación y Constitución Orgánica.* https://www.desenredando.org/public/libros/1992/agenda/AgendaDeInvestigacion-1.0.0.pdf

Lavell, A. (1994). *Al Norte del Rio Grande.* La Red.

Lewis, J. (1999). *Development in disaster-prone places: Studies of vulnerability.* IT Publications (Practical Action).

Lewis, J., & Kelman, I. (2012). The good, the bad, and the ugly: Disaster risk reduction (DRR) versus disaster risk creation (DRC). *PLOS Currents, 4,* e4f8d4eaec6af8. https://doi.org/10.1371/4f8d4eaec6af8

Lightman, A. (2013). *The accidental universe.* Vintage.

Lundgren, M., & Strandh, V. (2022). Navigating a double burden – floods and social vulnerability in local communities in rural Mozambique. *International Journal of Disaster Risk Reduction, 77,* 103023. https://doi.org/10.1016/j.ijdrr.2022.103023

Mansilla, E. (Ed.). (1996). *Desastres: Modelo para Armar.* La Red. https://www.desenredando.org/public/libros/1996/dma/DesastresModeloParaArmar-1.0.0.pdf

Maskrey, A. (1989). *Disaster mitigation: A community approach.* Oxfam. https://policy-practice.oxfam.org/resources/disaster-mitigation-a-community-based-approach-121119/

Maskrey, A. (Ed.). (1993). *Los Desastres No Son Naturales.* La Red. http://www.desenredando.org/public/libros/1993/ldnsn/LosDesastresNoSonNaturales-1.0.0.pdf

Maskrey, A. (1996). Prefacio a la Edición en Español (Preface to the Spanish edition). In P. Blaikie, T. Cannon, I. Davis, & B. Wisner (Eds.), *El Entorno Social, Político y Económico de los Desastres* (pp. 3–8). La Red. https://www.desenredando.org/public/libros/1996/vesped/

Mbaegbu, C. C. A. (2012). *Hermeneutics of god in Igbo ontology.* Fab.

Mercer, J. (2010). Disaster risk reduction or climate change adaptation: Are we reinventing the wheel? *Journal of International Development, 22*(2), 247–264. https://doi.org/10.1002/jid.1677

Michellier, C., Pigeon, P., Paillet, A., Theodore Trefon, T., Dewitte, O., & Kervyn, F. (2020). The challenging place of natural hazards in disaster risk reduction conceptual models: Insights from central Africa and the European Alps. *International Journal of Disaster Risk Science, 11,* 316–332. https://doi.org/10.1007/s13753-020-00273-y

O'Keefe, P., Westgate, K., & Wisner, B. (1976). Taking the 'natural' out of 'natural disaster'. *Nature, 260*(5552), 566–567.

Oliver-Smith, A., Alcántara-Ayala, I., Burton, I., & Lavell, A. (2016). *Forensic investigations of disasters (FORIN): A conceptual framework and guide to research (IRDR FORIN publication no. 2).* Integrated Research on Disaster Risk.

O'Mathúna, D. (2018). Christian theology and disasters: Where is god in all this? In D. O'Mathúna, V. Dranseika, & B. Gordijn (Eds.), *Disasters: Core concepts and ethical theories* (pp. 27–42). Springer.

Provalis. (2023). https://provalisresearch.com/

Sandoval Henríquez, V. (2017). *The progression of vulnerability: A multi-scalar perspective on disasters, the case of Chaitén, Chile* [Doctoral dissertation, The Bartlett Development Planning Unit, University College London]. https://discovery.ucl.ac.uk/id/eprint/1561339/

Stiekes, G., Talbert, L., & Casillas, K. (2017, September 11). Natural disaster and the new testament theology. *Theology in 3D.* https://seminary.bju.edu/theology-in-3d/natural-disasters-and-new-testament-theology/

Turner, B. L., Kasperson, R. E., Matson, P. A., McCarthy, J. J., Corell, R. W., Christensen, L., Eckley, N., Kasperson, J. X., Luers, A., Martello, M. L., Polsky, C., Pulsipher, A., & Schiller, A. (2003). A framework for vulnerability analysis in sustainability science. *Proceedings of the National Academy of Sciences, 100*(14), 8074–8079. https://doi.org/10.1073/pnas.1231335100

UNDRR. (2022). *The early engagement of the United Nations in disaster risk reduction (1970–2000): A brief history.* United Nations Office for Disaster Risk Reduction.

Watts, M., & Bohle, H. G. (1993). The space of vulnerability: The causal structure of hunger and famine. *Progress in Human Geography, 17*(1), 43–67. https://doi.org/10.1177/030913259301700103

White, G. (1945). *Human adjustment to floods* [Doctoral dissertation, Department of Geography, University of Chicago]. https://biotech.law.lsu.edu/climate/docs/Human_Adj_Floods_White.pdf

Wilder, J. (2018). *Operationalizing the pressure and release theoretical framework using risk ratio analysis to measure vulnerability and predict risk from natural hazards in the Tampa, FL Metropolitan area* [Doctoral dissertation, University of South Florida]. https://digitalcommons.usf.edu/cgi/viewcontent.cgi?article=8442&context=etd

Wisner, B. (2016, August 31). Vulnerability as concept, model, metric, and tool. *Oxford Research Encyclopedia of Natural Hazard Science.* Retrieved January 25, 2024, from https://oxfordre.com/naturalhazardscience/view/10.1093/acrefore/9780199389407.001.0001/acrefore-9780199389407-e-25

Wisner, B., & Alcántara-Ayala, I. (2023). Revisiting frameworks: Have they helped us reduce disaster risk? *Jàmbá: Journal of Disaster Risk Studies, 15*(1), a1491. https://doi.org/10.4102/jamba.v15i1.1491

Wisner, B., Blaikie, P., Cannon, T., & Davis, I. (2004). *At risk: Natural hazards, people's vulnerability and disasters* (2nd ed.). Routledge.

Wisner, B., Gaillard, JC., & Kelman, I. (2012a). Framing disaster: Theories and stories seeking to understand hazards, vulnerability and risk. In B. Wisner, JC. Gaillard, & I. Kelman (Eds.), *The Routledge handbook of hazards and disaster risk reduction* (pp. 18–34). Routledge.

Wisner, B., Gaillard, JC., & Kelman, I. (Eds.). (2012b). *The Routledge handbook of hazards and disaster risk reduction.* Routledge.

Wolbers, J., Kuipers, S., & Boin, A. (2021). A systematic review of 20 years of crisis and disaster research: Trends and progress. *Risk, Hazards & Crisis in Public Policy, 12*(4), 374–392. https://doi.org/10.1002/rhc3.12244

Yapa, L., & Wisner, B. (Eds.). (1995). Questioning development: Growth? Destruction? Sustainability? Theme issue. *GeoJounal, 35*(2), 99–104. https://doi.org/10.1007/BF00814056

Zimmerman, E. (1933). *World resources and industries: A functional appraisal of the availability of agricultural and industrial resources.* Harper and Brothers.

Chapter Three
Abilities

This chapter addresses people's abilities: how they tackle the disaster process at various stages and what influences abilities and may limit them. They will be examined from the point of view of the use and abuse of power in its socio-cultural, institutional, political, economic, and coercive forms. Discussion of people's ability to deal with hazards and disasters mirrors a broader debate in the social sciences about structure and agency. Arguing for human agency, we will present mounting evidence that people are not helpless. We will overview the strategies they deploy and see that these are grounded in the strength and diversity of everyday livelihoods and abilities (Wisner, 1993). Furthermore, we will suggest that these strategies reflect resistance to standardized and normalizing approaches to reducing disaster risk.

Abilities Beyond and Despite Disaster

In Chapter 1, we examined how the concept of vulnerability has allowed a movement away from naturalistic explanations of disasters (Wisner et al., 1976). Unpacking vulnerability has unlocked a new understanding of disaster that shifts attention from focusing solely on natural hazards (Wisner, 2016). Over the past 50 years, this understanding has gained considerable traction among scholars and practitioners and, to some extent, within policy circles. It underpins a discourse on the *construction* of disaster risk that reaches beyond the language of disaster risk *reduction* (Lewis & Kelman, 2012; Wisner, 2020). The discourse on vulnerability challenges those individuals and institutions with power to address the root causes of disasters and prevent the construction of new risks. However exposed and vulnerable to hazards ordinary people may be, they nevertheless

DOI: 10.4324/9781003292814-3

possess abilities to resist, cope with, adapt to, and recover from hazard impacts and to pursue individual and collective life goals despite disaster risk. Ability is sometimes called "capacity" with both the singular and the plural of each word employed.

These 50 years have witnessed uneven practical consequences of the shift in theory from a hazard-focus approach to an approach based on ability and vulnerability. In Europe and North America and more affluent countries in Asia and the Pacific as well as Latin America, social security, health care, and other social protection institutions were established and developed over the past century. Such social investment and public policy nourished local abilities. Africa and other less affluent world regions lacked effective stimulation of abilities. Various economic crises and sovereign debt management regimes eroded and weakened social protection (Pickety, 2014). From 2020, COVID-19 pandemic and post-pandemic changes in public finance and an uptick in military expenditure and conflict exacerbated pressure on social protection. The result is that more and more people have been forced to call upon their own abilities to maintain at least a precarious livelihood, whilst state support for local abilities dwindled.

Students of disaster and development have been slow to recognize the significance of ordinary people's abilities in justifying and implementing external interventions devised to "assist" vulnerable people in dealing with natural hazards. The portrayal of those affected by hazards as "helpless victims" has supported the goals and activities of international organizations engaged in human development, disaster risk reduction, and humanitarian response (Gibson, 2019). The image of people as helpless stirred concern among local organizations that work alongside ordinary people could undermine efforts to strengthen livelihoods and reinforce abilities.

In their book *Rising from the Ashes*, Anderson and Woodrow (1989) called for recognition of people's abilities. Since then, the word "capacity" has been used as a synonym for "ability". The singular form is used in the wider field of human development (Few et al., 2015; Gaillard et al., 2019). "Capacities" and "capacity" have often been used interchangeably, especially in the analysis and mapping of hazard, vulnerability, and capacity, or, as the professional jargon puts it, HVCA (Bhatt et al., 2011).

In this book, we will use "capacity" and "ability" interchangeably, although there may be a subtle difference. This difference involves the presence or absence of evidence that skills, knowledge, states of mind, and emotions constituting capacities have been performed or manifested in the past, although in the present, they may be dormant. If there is evidence of past mobilization of capacity and its performance, then the use of the term "ability" is warranted. Dormant capacities are the same as abilities. Theoretically, possible "capacities" latent in human beings and the human condition with no history of performance do not rise to the level of "abilities" and remain "capacities". One might nonetheless say that to the extent that capacities available to confront hazards are always dormant in daily life, they should always be considered abilities.

On Structure and Agency/Power and Resistance

Vulnerability is a situation that reflects how power and resources are shared within society, as we saw in Chapter 1. What is the respective role of vulnerability and ability in shaping people's behavior(s)? We need to ask whether or not there are mutually dependent interactions between vulnerability and ability. These questions are typically answered in two ways. The traditions of structuralism suggest that people's behaviors are constrained by structural forces beyond their reach. These are vulnerability's "root causes" (Wisner et al., 2004, 2012). This way of thinking leads to a consideration of unequal power relations as a factor preventing people from exerting their ability to deal with hazards and disasters. By contrast, existentialist traditions have argued that people are always able to express their free will and that the social fabric results from individual decisions. In such a context, people's ability to deal with hazards and disasters would be expressed as acts of self-determination.

In structuralist and existentialist traditions, vulnerability and ability are central concepts in theories underpinning radically different interpretations of disasters. Through a structuralist lens, the unequal distribution of power in society exposes people to hazards and disasters. From an existentialist perspective, people's decisions and their own deliberate choices expose them. In these ontologically distinct perspectives, reducing disaster risk is either a matter of addressing the root causes of vulnerability or of enhancing people's abilities through incentives for changed behavior and access to the resources to do so. We reject the either/or, arguing that reality is more complicated. Power, as Foucault would argue (1976), is a diffuse form of control consciously or unconsciously guided, through multiple formal and informal channels. Power relationships are embedded within the social fabric and penetrate all dimensions of everyday life. The most vulnerable always display some abilities and are never totally helpless when facing disaster and its root causes. Likewise, the most privileged and seemingly least vulnerable people may nevertheless not exhibit much knowledge of their own vulnerability or use all available resources dealing with hazards and disasters. A case in point is the behavior of the very rich who reside in exclusive Newport Beach, California (Wisner, 1999).

These reflections bring to mind a similar discussion initiated by Giddens (1984) and Bourdieu (1986) about the relative and combined contributions of structure and agency in shaping people's everyday actions. For both Giddens (1984) and Bourdieu (1986), there is a dialectical relationship between structure and agency. Giddens uses the term "structure" and Bourdieu the word "habitus". Both concepts provide space for people's agency to emerge and abilities to be expressed. One is also reminded of the perspective Sen (1992) adopted when he used the concept of the capability to understand people's ability to sustain their needs and achieve goals in life. Sen presented another example of the dialectical relationship between structure and agency. In dealing with hazards and disasters, this dialectic suggests that vulnerability and ability are situated at two ends of the same continuum defined by structure and agency. Thus, the most vulnerable people may be those with the least ability and vice versa.

We argue that this is not necessarily the case and that both vulnerability and ability are integrated components of disaster risk. Indeed, the display of abilities can constitute a form of resistance to power and vulnerability imposed on people without their consent. In such a perspective, resistance does not necessarily require direct confrontation. People may "vote with their feet" and simply ignore official advice because, for whatever reason – for example, mistrust or local impediments to implementation. As Foucault (1976) famously stated, where there is power, there is always resistance. Resistance is rationally geared toward safety rather than irrational (Quarantelli & Dynes, 1977; Kendra & Wachtendorf, 2016). Similar patterns of cooperative, proactive behavior have been observed in the context of risk reduction to prevent disasters from occurring. As early as the 1980s, a number of practitioners started to emphasize that local people are creative and resourceful in preventing hazards and recovering from their impacts (Maskrey, 1987; Anderson & Woodrow, 1989).

Despite such wealth of empirical evidence, there have been few attempts to theorize people's abilities. Exceptions include Anderson and Woodrow's (1989) vulnerability/capacity matrices and subsequent iterations, including the "circle of abilities" (Figure 3.1).

The circle of abilities shows how local people may use six kinds of resources to weave together strategies to gain access to decision-making that affects their lives, solidarity with others, and sufficient economic power to provide safe housing and their other basic human needs as well as some financial reserves for contingencies. These six kinds of resources should not be essentialized as the diagram may suggest. In practice, they are fluid and evade any form of categorization. They lead to flexible actions and strategies

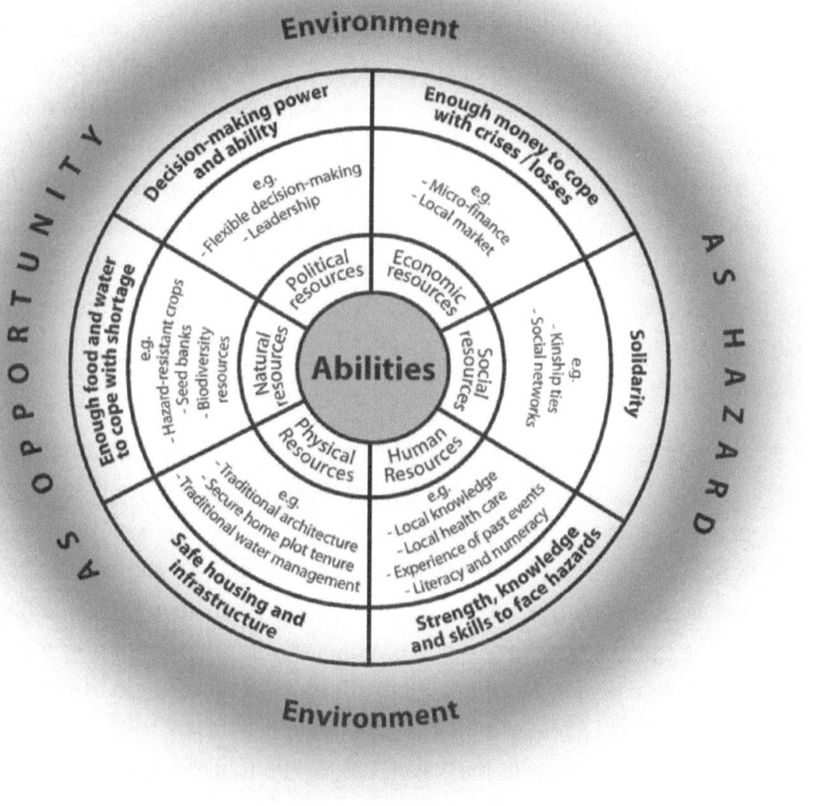

FIGURE 3.1 The circle of abilities

that build on both a fine-grained knowledge of the environment and past experience of similar situations (Wisner et al., 1977). They are often geared toward "coping", a concept related to capacity, widely and commonly found in disaster studies literature. Coping involves a limited exercise of abilities focused narrowly on the nature and scope of a particular hardship at a moment in time. For example, the focus might be providing immediate shelter from the elements when one's house is destroyed or finding food for the next day or two, rather than using abilities strategically to avoid such situations in the first place. Their actions are neither static nor rigid. They evolve, combine, and diversify. If conditions allow, they may transcend day-to-day coping and take on a more strategic form of capacity. Favorable conditions allowing the evolution beyond mere coping depend on the resources named in the circle of abilities.

The abilities deployed in facing hazards and disasters are a hybrid mix from various sources because people increase available opportunities by tapping different sources of knowledge and experience. For example, in Kenya, a parent may consult a local herbalist or spiritual healer as well as a Western-trained clinician when caring for a sick child. The sources may be traditional, loaned, supported by the latest Western technology, or any combination. The abilities on which people draw are both individual and collective. Each individual carries their own bundle of knowledge, resources, and skills but can use these in combination with those of relatives and neighbors to enhance the risk reduction outcome (Gaillard et al., 2019). There may also be non-local support, such as that provided by transnational networks, including remittances from migrant workers or abilities they have learned from others while abroad.

As evidence of local people's abilities mounted up, there was a growing recognition that governments and NGOs should build on these endogenous abilities rather than heavy-handedly impose exogenous solutions. The 1970s witnessed this recognition become political advocacy. The demand was that people should be able to control their own lives. People's abilities had been ignored and their voices neglected or silenced. They should now be at the forefront of any efforts geared toward improving their everyday life (Freire, 1970), including efforts to reduce disaster risk. Recognition of people's abilities underpinned the discourse on people's participation in development that built momentum in the 1970s and 1980s (Chambers, 1983). It ultimately trickled through to the alternative understanding of disasters in the late 1970s (Wisner et al., 1977) and 1980s (Anderson & Woodrow, 1989) in the form of vulnerability and capacity analysis.

Packaging Abilities

People with administrative, coercive, or economic power positions in any sector have a keen ear for local language describing plants and animals. Government officials will use vernacular terminology to package their desired or prohibited behaviors, such as contour ridging (officially desirable but subject to local skepticism – Blaut, 1959, 1960; Zimmerer, 1994), or the use of fire (officially condemned) by shifting cultivators (Carmenta et al., 2013). Many organizations want to identify possible plants and animal products they may exploit as commodities and may go as far as attempting to acquire legal patents, hence, owning their genes, a practice that has been labeled biopiracy (Shiva, 1999). The common practice of vulnerability and capacity analysis (VCA) in disaster studies and practice is another instance of the powerful appropriating local knowledge and abilities for their own purposes, scientizing it and emptying the original of its uniqueness (Agrawal, 1995).

Although it recognizes people's abilities and encourages people's participation in disaster risk reduction programs, the discourse and practice of VCA have largely been emptied of its collective resistance and joint power. In their place, one found a neoliberal emphasis on individual responsibility (Harvey, 2005; Gaillard, 2021). Rather than relying on collectively pooled and hybrid abilities, organizations have come to employ standardized, exogenous toolboxes to assess and strengthen people's abilities. These toolboxes draw on taxonomic categorizations of people's resources, skills, and knowledge with some minor tweaking to accommodate local realities (e.g., International Federation of Red Cross and Red Crescent Societies, 2007; Dazé et al., 2009; Ahmed et al., 2012). The categories have been identified through quantitative, often demographic, indicators. Standardized indicators appear as tick boxes wherever in the world the VCA is conducted. Bhatt (1998) warned that standard indicators and methods designed by outsiders are likely to be "filtering what she or he reads through the conceptual framework, assumptions, and values or her or his culture and, as a result, is creating false 'stories' that fit her or his expectations" (p. 71). Indeed, the use of VCA and similar methods to understand people's abilities rely on a Western desire to distinguish, sort, count, and categorize. Whereas such procedures and ways of thinking seem natural and common sense to Westerners and Western-trained researchers and policy makers, they may seem quite alien to people beyond the West (Gaillard, 2021).

VCA toolboxes usually entail activities that are to be facilitated in a predefined sequence, such as the elaboration of a timeline of past hazard events, then some scoring and ranking of hazards, then brainstorming matrices and problem trees to unpack vulnerabilities and capacities, and so on through the capacious toolbox. In the worst cases, a toolbox requires following a standard timeframe with preestablished daily or weekly programs of activities that mirror a project management approach. This approach is geared toward upward accountability and to meeting the expectations of funding agencies and donors. In fact, these initiatives are often labeled "capacity building" (Kaplan, 2000; Few et al., 2015) or "capacity development" (Capacity for Disaster Reduction Initiative, nd), thereby suggesting a transfer of external knowledge and capacities that goes against the very ethos of the discourse emphasizing people's own abilities (Eade, 2007). Bhatt and Pandya (2006, p. 1) found that ironically, "we" (outsiders) end up focusing "on building our capacity to reduce their [local people's] risks or our capacity to build their capacity to reduce risks" instead of enhancing people's own ability. This standardized approach proves ill-suited to understanding the often intangible way in which local people's abilities are embedded within local cultures and subcultures – ways that make them hidden transcripts (Scott, 1990) of disaster risk reduction.

Hidden Transcripts of Disaster Risk Reduction

We have seen that people's abilities to reduce disaster risk are embedded in culture. The abilities of the poorest and most vulnerable are not necessarily embedded in the mainstream culture of central, dominant, more powerful people; rather, it rests more often in a subculture (Anderson, 1965; Wenger & Weller, 1973; Wenger, 1978). The prefix "sub" can signal the hidden nature of the resources, skills, and knowledge that support people's abilities. Not only are these abilities out of step with the mainstream but they may also provide a mode of conscious or unconscious resistance to dominant forms of governing disaster risk. This is the same "sub" used in the word "subaltern" – denoting the colonized who are denied access to political, social, and economic power. Therefore, the "sub" in subculture is also political.

People's intrinsic abilities to deal with hazards and disasters can be read as a tacit and potential form of resistance. Scott (1990) suggests that those subjected to technocratic and normative forms of governance publicly accept domination but actively subvert formal and official initiatives, whether from the government or other stakeholders, "off-stage". As such, "resistance is virtually always a stratagem deployed by a weaker party in thwarting the claims of an institutional or class opponent who dominates the public exercise of power" (Scott, 1989, p. 52). These are, therefore, hidden transcripts of resistance. People publicly welcome normative disaster risk reduction initiatives from the government and other actors, such as NGOs, as a gesture of goodwill and means to leverage other forms of support (White, 1996; Grove, 2013). However, people's actual response to threats to their lives and livelihoods may differ from what formal plans and procedures mandate. For example, the response to earthquake, flood, or fire of people locked up behind bars in jails and prisons in the Philippines most often differ from what is recommended in the formal and rigid emergency response plans of the penitentiary administration (Gaillard, 2021). Nonetheless, the response of people in jails and prisons, which is grounded in cultural values inherent in society and everyday life in the Philippines and in abilities outside of jails and prisons, often proves successful in avoiding major losses.

Such outcomes in dealing with hazards and disasters remain hidden transcripts because, in the case of jails and prisons in the Philippines, it is not in the interest of those behind bars to disclose that they hold more power than they should. That is, they can organize their own evacuation in partnership with their guards. The authorities do not have any reason for acknowledging that their formal plans are of limited use either since, as Scott (1989, p. 49) suggests:

> it is also [. . .] rare that officials of the state wish to publicise the insubordination behind everyday resistance. To do so would be to admit that their policy is unpopular and, above all, to expose the tenuousness of their authority. [. . .] The nature of the acts themselves and the self-interested muteness of the antagonists thus conspire to create a kind of complicitous silence which may all but expunge everyday forms of resistance from the official record.

Recognition of people's abilities would undermine the very agenda of many non-state actors, especially NGOs, whose long-term existence relies on emphasizing the impact they make for "reducing disaster risk" through their formal initiatives, such as awareness campaigns, formal plans, and drills. The efforts of actors, such as NGOs, are nonetheless acknowledged and often welcomed by local people because doing so allows them to secure political allegiance and leverage support for other purposes (White, 1996) (see Box 3.1). In Scott's words, people's "tactical wisdom" remains hidden and deeply political (Scott, 1989, p. 57).

BOX 3.1 Songs of Confrontation and Resistance

In the aftermath of the 2010 flood in São Luiz do Paraitinga, Brazil, external agents, including the Armed Forces, and national and state agencies tried to control the situation of the poor and survivors (Marchezini, 2015a). The local inhabitants, who call themselves *luizenses*, maintain a lifestyle called *caipira* – typical of a traditional rural community. *Caipira* is considered a pejorative term when used by outsiders because its character is portrayed as uneducated, rural, based on family values, and rooted in tradition (Marchezini, 2015b). The *luizenses* celebrated the hidden abilities they possess by performing everyday cultural resistance in songs they sing during annual carnival festivities. One of these songs, composed in the aftermath of militarized flood response, used the verb "to confront":

> We made some songs, I remember one that I made. We confronted prejudice and the dictatorship, the Carnival procession will never contain it,

we confronted the flood in a dark night, taking water from stone and honey from sugar, in the hammering of reconstruction, tile by tile, heart by heart.

(Marchezini's field notes, December 2011)

Unpacking the song, hidden transcripts of people's abilities emerge. "We confronted the prejudice and the dictatorship" evokes the invigorating social energy or resistance generated during the first days after the flood, when the Armed Forces and the military police were relocated to the city and used violence directed at the *luizenses* to assert control. "We confronted the flood in a dark night" recalls the rescues based on the collective work of city residents who were also rafters. The rafters ended up having to rescue some of the military who were supposed to be the rescuers. After the flood, they became known as "rafting angels". "Taking water from stone and honey from sugar, in the hammering of the reconstruction, tile by tile, heart by heart" is a symbolic language that synthesizes the difficulties involved in the process of reconstructing the city when the spontaneous solidarity of civil society had ended (Marchezini, 2015b, p. 299), as sociologists of disaster agree it eventually must end (Gonzalez-Mendez et al., 2021).

Influences on Abilities and Their Limitation

Having looked closely at the range of skills and varieties of knowledge people utilize in dealing with hazards and disasters in their daily lives and how these abilities are expressed in specific cultural and institutional contexts, we turn now to consider what influences abilities and what might limit them. These influences include the exercise of power in its economic, political, socio-cultural, and administrative forms.

As we saw in Chapter 1, political regimes and their associated understanding of risk and its management are shaped by people, but people are in different positions of power. While some cooperate and manifest their normally hidden transcripts of resistance in situations of economic and political oppression, others remain in the shadows, combining their knowledge, experience, and resources to maximize their economic, social, and political benefits without engaging directly with the state or other holders of power. In other words, structure and agency are dynamically shaped by social power relations across temporal and spatial scales.

Political regimes affect the root causes of disaster vulnerability by influencing social and economic structures, such as the characteristics of power, wealth, and distribution of resources, as well as ideology. Regimes impact dynamic pressures, such as technological change, deforestation, commodity price fluctuations, housing markets, business cycles, and land use (Wisner et al., 2004; Wisner, 2016). They affect social protection policies, disaster risk construction, and management. They influence and are influenced by people's abilities. When analyzing this dialectical relationship, it is important to consider different temporal and spatial scales. Regimes' actions or inactions may encourage or discourage, even block, local abilities to deal with hazards and disasters. Abilities, as we have seen, are intimately enmeshed in the cultures of people-in-a-place. There are various organizations whose conscious goal is to reveal, protect, and conserve local knowledge and abilities, such as social movements and NGOs. Others, especially for-profit corporations, wish to exploit local abilities. An example of the latter is biopiracy – that is, corporate attempts to patent species, particularly with medicinal uses, revealed by local botanical knowledge. A detailed discussion of governance and the interventions by social movements and NGOs will reveal further ramifications of the relationship between structure and agency.

Ideology and Governance as Influences on Abilities

Galeano (1973) analyzed the geopolitics of colonial exploitation, providing examples of atrocities by intruders, such as Spaniards, Portuguese, French, Dutch, British, and North Americans, beginning as early as the 15th century. The rush for commodities, such as gold and silver, uprooted Indigenous peoples from their territories, impacting their living conditions and the symbolic meanings used to shape their territories (Haesbaert, 2004). These intrusions and conquests affected living and non-living elements of the human environment, all of which were endowed with meanings as well as a pragmatic utility: water in the form of rivers, creeks, waterfalls; plants in plains and forests; sacred trees, sacred animals, and sacred spaces, including hillsides, groves, stone circles, and burial mounds; and much more.

Galeano (1973) analyzed the demarcation of boundaries and, hence, the extent of states' political power and their projection of power over landscapes. According to Van Houtum (2005), boundaries and borders are different. A boundary is a geopolitical projection of power. A border is a socio-territorial construct produced by social practices (Bourdieu, 1991) and represents differences in space. For Van Houtum (2005, p. 677), there is a need to discuss the ethics of socio-spatial borders: "to what extent is it morally just to protect ourselves and thereby deny the liberty of access to others?" This question of "access" defines a group's abilities, especially those who depend on a place for subsistence.

Building a "home" – a sense of place and the act and experience of "dwelling" – is one of the basic abilities required to live a fully human existence (Bachelard, 1964). As contemporary controversies over migration remind us, the "right to stay" may be contested in different forms, using physical and/or symbolic violences.

The macro-forces that shape borders as well as local abilities include ideologies. Ideologies may be shared by diverse vehicles of state power. Authoritarianism, for instance, has been normalized in territories where the state apparatus ranges from democracy to autocracy. The 2022 edition of the annual global report, *Freedom in the World*, noted that political rights and civil liberties have declined worldwide, indicating that autocracy could overtake democracy as the governance model (Repucci & Slipowitz, 2022, p. 1). The global report highlights military coups and power grabs as more common in 2021 than in any of the previous ten years. The authors showcase several examples of autocracies worldwide. The opening of this curse can nurture new forms of necropolitics that suppress local abilities to resist and cope with hazards. It can create unequal access to resources needed to face disaster risks and increase exposure to risks of harm, loss, injury, or even death for social groups that are stigmatized or made vulnerable (Mbembe, 2019; Sousa-Santos, 2021).

In 2019, the first year of Brazilian President Jair Bolsonaro's time in office, the new government expelled Cuban doctors from 3,600 municipalities, where they would have been even more necessary during the COVID-19 pandemic (Sousa-Santos, 2021). From the beginning of the COVID-19 pandemic, the Brazilian president gave speeches and provoked actions that opposed recommendations from medical science for reducing risk (de Freitas et al., 2020).

Governance systems determine the scope of social protection policies, including access to income security, health care, and decent work. Implemented well, social protection can provide space for local abilities to flourish. The *World Social Protection Report 2020–2022* found that 53% of the global population (over 4.1 billion people) did not have income security (International Labour Organization [ILO], 2021). Such inequalities manifest at the regional level, and differences appear in gender, age, urban and rural residence, and types of social protection. Worldwide, most unemployed workers (81.4%) are not covered by unemployment benefits. One in three people with a physical or mental impairment receives a benefit. Only 45% of women with infants receive a cash maternity benefit, while only one in four children (26.4%) receives a social protection benefit. These situations are heightened by the intersection of social variables, making some groups more vulnerable than others.

Corporations and Social Movements

Our abilities' limits are shaped not only by political systems but also by organizations, for-profit and nonprofit, and social movements. Corporations have aimed to influence government policy for centuries, with the magnitude and tactics of influence continuing at extreme levels in the 21st century. Nyberg (2021) lists several tools at corporations' disposal:

- Funding political campaigns.
- Partnering with think tanks to shape policy agendas.
- Hiring lobbyists to strengthen relationships with policy makers.
- Creating their own public relations departments.
- Former employees securing political and public service positions once retired ("the revolving door").

The tremendous power of corporations across areas and populations has led some authors to use the term "corporate sovereignty" (Kapferer, 2004) (Box 3.2). Clapp (2021) examined the implications of corporate sovereignty in the global seed and agrochemical industry, where four large firms:

> influence the types of seeds farmers plant, what crops are grown, what breeds of livestock are raised and in what types of facilities, working conditions for food system workers, and the types and prices of food items that appear on grocery store shelves.
>
> (p. 404)

Such concentration of economic power squeezes out the small, independent producer, their local knowledge of the land and its non-human life, as well as their local abilities. Corporate oligopoly as well as vertical and horizontal integration construct disaster risk in a number of ways. First, it ties food security to a narrow genetic base. This was a cause of the potato's vulnerability to blight in Ireland and its infamous famine (1844–1848) (Wisner et al., 2004, pp. 178–180). Second, the food system becomes dependent on a complex commodity chain that can fail, as the Russia's full-scale invasion of Ukraine in 2022 and the COVID-19 pandemic showed. Finally, the pressure the corporate system puts on wild animals through expansion into newly exploited biomes increases the risk of epizootic disease outbreaks that may spread to humans (see Chapter 4).

BOX 3.2 Corporate Lobbying in the Food System

In 2009, the top four agro-industrial firms' share of the global market was 54% for seeds and 53% for agrochemicals. By 2018, the four large firms controlled 70% of the global pesticides and 60% of the global seed market. The investments of these mega firms in lobbying were substantial. "In 2019 . . . Corteva Agriscience spent over US$3 million and BASF and Syngenta each spent over US$1 million on lobbying activities in the USA. Bayer AG spent US$9 million in the same year . . . Bayer and BASF each also spent more than €3 million (USD 3.3 million) lobbying the European Union in 2019; while Syngenta spent more than €1.5 million and Corteva nearly €1 million that same year" (Clapp, 2021, p. 406). The scale of such corporate influence seriously affects the ability of ordinary people to protect themselves from hazards arising in the agro-industrial system, the hazards of environmental pollution, and biodiversity erosion.

Not only is the ability of ordinary people potentially blocked or reduced but, also, their foresight and understanding of risk may be obscured or confused by social media and other means of communication. People can be manipulated by false news and information made increasingly compelling by sophisticated technology like generative artificial intelligence. Horkheimer and Adorno (1947) coined the term "cultural industry" in the 1940s describing how culture may be co-opted by corporations and transformed into a source of profit through mass consumption. They could not have imagined that such cultural-industrial production would grow to constitute at least 63% of global online content (International Telecommunication Union [ITU], 2021). This digital content is influenced by corporate sovereignty since, as Nyberg (2021, pp. 11–12) puts it:

> media outlets are shifting their business models to provide free news content, which makes them increasingly dependent on advertising. (. . .) Through advertising, corporations buy not only a platform for communicating political messages but also increasingly favourable treatment from the media.

Of course, the internet nevertheless provides sources that can help people to demystify and debunk false news and claims. There is valuable content and important access to tools that can enhance the ability to foresee and reduce vulnerability. However, the digital divide limits the use of these ability-enhancing resources (Box 3.3).

BOX 3.3 The Digital Divide

The percentage of people using the internet in 2021 increased by 17% compared to 2019: "Internet penetration increased more than 20% on average in Africa, in Asia and the Pacific, and in the UN-designated Least Developed Countries (LDCs)" (ITU, 2021, p. iii). 2.9 billion people, 96% of whom live in less affluent countries, remain without internet access. The digital divide occurs between generations. 71% of the world's population aged 15–24 uses the internet, compared to 57% of all other age groups. There is also unequal access by gender. 62% of men use the internet compared to 57% of women (ITU, 2021). Moreover, although 95% of the world's population lives within a broadband network's range, there are zones without coverage: 30% of the rural population in Africa lacks broadband network access (ITU, 2021).

Thus, many of the factors which shape local abilities are in the hands of corporations. Social movements maintain positions in this field of power that are not determined solely by economic and financial capital, with social, scientific, political, and cultural influences too (Bourdieu, 1991). Social movements gather individuals in a collective effort to fight against institutions and situations that block or diminish their abilities. For instance, in the 1980s, a social movement called attention to the disproportionate dumping of toxic waste in urban or peri-urban areas where people of color live. This social movement embraced the idea of environmental justice and the principle that "all people and communities are entitled to equal protection of environmental, energy, health, employment, education, housing, transportation, and civil rights laws and regulation" (Bullard, 2021, p. 244).

Two decades later, there has been a proliferation of social movements, ranging from far-right to far-left, from anti-science to science-without-critique, and from those denying human-caused climate change to those thinking that human-caused climate change leads to all problems – as well as those supporting land and agrarian reform, fighting hunger, protecting biodiversity, tackling human slavery and trafficking, and much more. These movements have different funding mechanisms, actors, and spheres of influence, including linkages with other movements. They are not treated equally by governments, media, corporations, or each other, frequently facing various forms of repression (Earl et al., 2022). Despite their differences, or perhaps because of it, social movements and NGOs are important for promoting awareness of threats to abilities and supporting participation in campaigns to protect abilities (Martinez-Alier et al., 2016; Wisner et al., 2020).

Types and Spheres of Participation

Participation is evoked by many international agendas, such as the Sendai Framework for Disaster Risk Reduction (SFDRR), the Paris Agreement, the Sustainable Development Goals (SDGs), the Addis Ababa Action Agenda, and Aarhus Convention (the UNECE Convention on Access to Information, Public Participation in Decision-making and Access to Justice in Environmental Matters). The latter includes a protocol on people's right to information about pollution (UNECE, 2023). International organizations responsible for these agendas do not intend a "strong" form of participation (Wisner, 1988); rather, their calls for "participation" do not demand systemic changes that would prevent future exclusion, discrimination, or unjust distributions of benefit and harm.

In George Orwell's novel *Animal Farm* from 1945, "all animals are equal, but some are more equal than others". Likewise, "participants" do not have equal positions in the field of power defined by their access to economic, political, scientific, and cultural capital (Bourdieu, 1986). For instance, it can be life-threatening for certain groups or individuals to demand information about safety or to claim participation in policy debates about safety in autocratic regimes. It can be impossible to participate in public debates when one has a daily struggle to obtain food or to cover the basic cost of living. It can be challenging to be heard when one does not have access to the internet, when the local press is funded by a corporation that creates disaster risks in that locality, or when one lives in a territory under paramilitary or military control.

In some situations, institutions will ask for a person's suggestions on a specific issue, but there is no guarantee that such input will go into the final decision. One may also face situations in which an organization will provide a lot of recommendations for reducing disaster risks without listening to others' concerns. Organizations use different forms of participation (Table 3.1) with various social groups and for different purposes. These types of participation strongly influence whether abilities are amplified or reduced.

Participation is an ideal to which many institutions genuinely aspire or endorse cynically to serve their own interests because it increases their political, economic, and social standing in relation to funders and public opinion. Unfortunately, there is a deep gap between the ideas shared during sparkly moments of participation in group discussions

TABLE 3.1 Types of participation

Level	Power relationships	Type of participation
Active	Participants set their own agendas. Learning occurs through the negotiation of ways to carry out agendas and actions in collaboration, and power may shift depending on the negotiations.	Co-acting
	Participants use different forms of knowledge to integrate new understandings. They define common agendas and share responsibilities within existing institutional and social settings and constraints.	Co-creating
Passive	One group takes the initiative and power for enticing other groups to act. They may set jointly issues, such as agenda and priorities.	Enticing
	One group (often the government) collects information from different groups but decides on the final project.	Consulting
	Information is usually just formal in a one-way flow. It uses technical language, and people often feel intimidated to express their views.	Informing
	The will of one group is imposed upon others. People cannot give opinions nor defend their interests.	Coercing

Source: Dyball et al. (2009)

or "trainings" and the implementation of measures to improve the well-being of those suffering in disasters and reinforce local people's abilities.

Distortions and Abuses of Participation

At the local scale of a primary residence and livelihood, "participation" has usually connoted the possibility of an amicable, cooperative process of deliberation and reflection leading to the local action, reflection, and further action that Wisner (1988) called "transformative" participation. Even at that local scale, there will be differences in material interests and conflicting goals. When one turns to large-scale human institutions, conflict and competition can increase. Public and private organizations compete, as well as NGOs, scientists, research institutes, United Nations offices, and others. Lund (2006) reminds us that public agencies compete for financial and human resources, emphasizing that the state is not a monolithic entity; rather, it is composed of such competing organizations. Competition may be for positions of influence over policy decisions ("a seat at the table") and for media visibility and status. The dynamics of such jockeying for position and power can eliminate space for local abilities to arise and flourish. As the saying goes, "When elephants fight, the grass is trampled".

Whether local abilities are recognized depends as well on the ability of the UN, government and non-governmental analysts, and decision-makers to perceive and understand local hazards. Organizations tend to frame their definitions of disaster risks and tailor their problem focus according to the solutions they have in hand (Dombrowsky, 1998). Without binding agreements that change the status quo, the agents involved in the field of power will keep updating their buzzwords and discourses, sweeping root causes of disaster under the rug. Participation, in this case, becomes routine and formalized. Such a state of affairs is not conducive to the discovery and appreciation of highly localized hazards, their highly localized impacts, and their highly localized meanings. Failing to notice these highly localized contexts, others will also miss highly localized abilities that seek to prevent and reduce hazard impacts and that guide recovery from them.

Within state and non-state organizations, those who control the means of policy implementation – such as budget, staff, protocols, and procedures – are not typically on the frontline with those who suffer everyday risks and disasters. The bureaucrat who invites local people "to participate" only rarely extends the invitation to develop and review the budget. On the contrary, the most common invitations involve only the lower part of the ladder of participation (Table 3.1), what Wisner (1988) called "instrumental" participation.

Yet around the world, there are still many examples of higher-order participation that achieves "co-creating" and even "co-acting" (Table 3.1). Much involves the work of social movements (Gibson, 2019; Gibson & Wisner, 2016). They fight for the progression of safety (Chapter 10), using protests, strikes, and other forms of nonviolent resistance against the construction of risks that negate local abilities (Chapter 9). Such actions by social movements include a focus on everyday risks people face and people's abilities in the face of such quotidian insults. Social movements can provide the base and energy to challenge the root causes of disaster vulnerability at the scale of state power. Chenoweth (2020) lists 628 mass campaigns between 1900 and 2019 seeking to remove national leadership from power or to create territorial independence by secession or expulsion of colonial powers, of which more than half (325) are said to have used nonviolent civil disobedience.

As highlighted earlier in this chapter, it may be that some of the most effective and sustainable initiatives to both resist the construction of risk and foster meaningful reduction of existing risk are and will remain hidden transcripts, in Scott's (1990) sense. That is, to be effective and sustainable, many of these initiatives that build upon people's organic abilities have to remain outside of the formal spaces of governance created by national and international policies as well as participatory processes encouraged by NGOs and their partners. As per our earlier vignette, if jails and prisons in the Philippines do not suffer major disasters, then it is because the intrinsic abilities of the detainees remain "offstage". Formal recognition of these abilities in policy would lead to their denial and retaliation for those detained (Gaillard, 2021).

Conclusions

Of the more than 7,000 languages spoken today, if a proto-Indo-European category is accepted, then perhaps over 440 languages spoken by about half the world's population have such roots. These languages have formulated the contemporary, dominant discourse of disaster. This discourse may or may not make sense beyond the expressions stemming from these roots. It has been imposed all over the world to the detriment of other understandings of the world, hazards, vulnerabilities, risks, disasters, and abilities. Recognizing and respecting the abilities of ordinary people residing around the world, who speak those other thousands of other languages, requires challenging power relations between those who impose top-down, standardized practices of disaster risk reduction. In this way, unheard voices and abilities can emerge and be heard. We can then enter into a respectful exchange with everyone based on mutual trust.

Yet trust may be hard to establish because, for centuries, macro-structural forces unleashed by the dominant minority have "managed" and sometimes reduced local abilities without asking for permission or feedback. Autocracies and other forms of state power have come and gone. In each, those with power attempted to control those without power. In contemporary times, local abilities continue to be diminished or blocked by a lack of social protection and by corporate sovereignty. Pushing back dominant elites are some NGOs and social movements that attempt to support and advocate for local participation. Some of them do so without questioning or challenging the macro-forces that assail and distort local abilities. Some of them end up perpetuating and enhancing the forces they claim to oppose.

A possible obstacle to gaining people's trust and cooperating with them is tension between recognizing that abilities are embedded within people's everyday life compared to the outsider's potential desire to find common features and patterns among abilities – how they arise, change, are challenged, are shared, and are reinforced. The human brain looks for patterns. Short of highly skilled meditation or deep anesthesia, the brain cannot necessarily be quietened, even in unconsciousness, such as sleep or being knocked out. But the quest for pattern need not replicate a Western taxonomic crusade. Other locally sensitive ways of finding common ground and patterns are possible and necessary. This task requires a very fine understanding of local contexts in both their tangible and intangible expressions. It is a task that, in theory, those at risk are often in the best position to lead.

References Cited

Agrawal, A. (1995). Dismantling the divide between indigenous and scientific knowledge. *Development and Change, 26*(3), 413–439. https://doi.org/10.1111/j.1467-7660.1995.tb00560.x

Ahmed, I., Fuenfgeld, H., & McEvoy, D. (2012). *Integrated disaster risk reduction and climate change: Participatory capacity and vulnerability analysis (PCVA) toolkit.* Oxfam Australia. https://unfccc.int/files/adaptation/cancun_adaptation_framework/adaptation_committee/application/pdf/pcva_toolkit_oxfam_australia.pdf

Anderson, M., & Woodrow, P. (1989). *Rising from the ashes: Development strategies in times of disasters.* Westview Press.

Anderson, W. (1965). *Some observations on a disaster subculture: The organizational response of Cincinnati, Ohio, to the 1964 flood. Research note 6, Columbus disaster research center.* The Ohio State University. http://udspace.udel.edu/handle/19716/1242

Bachelard, G. (1964). *The poetics of space.* Penguin.

Bhatt, M. (1998). Can vulnerability be understood? In J. Twigg & M. Bhatt (Eds.), *Understanding vulnerability: South Asian perspectives.* (pp. 68–77). Intermediate Technology Publications.

Bhatt, M., Chauhan, V., Patra, J., & Pathak, V. (2011). HVCA: Thinking beyond projects. *Southasiadisasters.net, 79,* 6–16. https://www.preventionweb.net/files/21713_21713participatoryhazardvulnerabili.pdf

Bhatt, M., & Pandya, M. (2006, February 13–15). *Rethinking capacity for disaster development risk reduction: Lessons from bottom up.* Global Future Search Meeting – Rethinking Capacity Development for Disaster Risk Reduction: Action 2005–2015, Geneva.

Blaut, J. (1959). A study of cultural determinants of soil erosion and conservation in the Blue Mountains of Jamaica. *Social and Economic Studies, 8*(4), 403–420. https://www.jstor.org/stable/27851232

Blaut, J. (1960, September 14–20). *The nature and effects of shifting agriculture.* Symposium on the Impact of Man on Humid Tropics Vegetation, Unesco South-East Asia Science Cooperation Office, Canberra, 185–198. https://portals.iucn.org/library/node/24552

Bourdieu, P. (1986). The forms of capital. In J. Richardson (Ed.), *Handbook of theory of research for the sociology of education* (pp. 241–258). Greenwood Press.

Bourdieu, P. (1991). *Language and symbolic power.* Polity Press.

Bullard, R. (2021). Environmental justice once footnote, now headline. *Harvard Environmental Law Review, 45*(2), 243–248.

Capacity for Disaster Reduction Initiative. (n.d.). *Basics of capacity development for disaster risk reduction.* Capacity for Disaster Reduction Initiative.

Carmenta, R., Vermeylen, S., Parry, L., & Barlow, J. (2013). Shifting cultivation and fire policy: Insights from the Brazilian Amazon. *Human Ecology, 41*(4), 603–614. https://doi.org/10.1007/s10745-013-9600-1

Chambers, R. (1983). *Rural development: Putting the last first.* Longman.

Chenoweth, E. (2020). The future of nonviolent resistance. *Journal of Democracy, 31*(3), 69–84.

Clapp, J. (2021). The problem with growing corporate concentration and power in the global food system. *Nature Food, 2,* 404–408. https://doi.org/10.1038/s43016-021-00297-7

Dazé, A., Ambrose, K., & Ehrhart, C. (2009). *Climate vulnerability and capacity analysis handbook.* CARE International.

Dombrowsky, W. (1998). Again and again: Is a disaster what we call a 'disaster'? In E. Quarantelli (Ed.), *What is a disaster? Perspectives on the question* (pp. 19–30). Routledge.

Dyball, R., Brown, V., & Keen, M. (2009). Towards ustainability: Five strands of social learning. In A. Wals (Ed.), *Social learning towards a sustainable world: Principles, perspectives, and praxis* (pp. 181–194). Wageningen Academic Publishers.

Eade, D. (2007). Capacity building: Who builds whose capacity? *Development in Practice, 17*(4–5), 630–639.

Earl, J., Maher, T., & Pan, J. (2022). The digital repression of social movements, protest, and activism: A synthetic review. *Science Advances, 8*(10). https://doi.org/10.1126/sciadv.abl8198

Few, R., Scott, Z., Wooster, K., Avila, M., Tarazona, M., & Thomson, A. (2015). *Strategic research into national and local capacity building for DRM: Synthesis report.* International Federation of Red Cross and Red Crescent Societies.

Foucault, M. (1976). *Histoire de la sexualité I: la volonté de savoir.* Gallimard.

Freire, P. (1970). *Pedagogy of the oppressed.* Bloomsbury.

Freitas, C. M. de, Silva, I. V. de M., & Cidade, N. da C. (2020). COVID-19 as a global disaster: Challenges to risk governance and social vulnerability in Brazil. *Ambiente & Sociedade, 23,* e0115. https://doi.org/10.1590/1809-4422asoc20200115vu2020L3ID

Gaillard, JC. (2021). *The invention of disaster: Power and knowledge in discourses on hazard and vulnerability.* Routledge.

Gaillard, JC., Cadag, J., & Rampengan, M. (2019). People's capacities in facing hazards and disasters: An overview. *Natural Hazards, 95*(3), 863–876.

Galeano, E. (1973). *Open veins of Latin America: Five centuries of the pillage of a continent.* Monthly Review Press.

Gibson, T. (2019). *Making aid agencies work: Reconnecting INGOs with the people they serve.* Emerald.

Gibson, T., & Wisner, B. (2016). 'Let's talk about you . . .': Opening space for local experience, action and learning in disaster risk reduction. *Disaster Prevention and Management, 25*(5), 664–684.

Giddens, A. (1984). *The constitution of society.* Polity Press.

Gonzalez-Mendez, R., Díaz, M., & Triberti, S. (2021). Volunteers' compassion fatigue, compassion satisfaction, and post-traumatic growth during the SARS-CoV-2 lockdown in Spain:

References Cited

Self-compassion and self-determination as predictors. *PLoS One, 16*(9). https://doi.org/10.1371/journal.pone.0256854

Grove, K. (2013). Hidden transcripts of resilience: Power and politics in Jamaican disaster management. *Resilience, 1*(3), 193–209.

Haesbaert, R. (2004). *O mito da desterritorialização: do fim dos territórios à multiterritorialidade.* Bertrand Brasil.

Harvey, D. (2005). *A brief history of neoliberalism.* Oxford University Press.

Horkheimer, M., & Adorno, T. (1947). *Dialektik der Aufklärung. Philosophische Fragmente.* Querido Verlag N.V.

ILO (International Labour Organization). (2021). *World social protection report 2020–22: Social protection at the crossroads – in pursuit of a better future.* International Labour Organization.

International Federation of Red Cross and Red Crescent Societies. (2007). *VCA toolbox with reference sheets.* International Federation of Red Cross and Red Crescent Societies.

ITU (International Telecommunications Union). (2021). *Measuring digital development. Facts and figures 2021.* ITU. https://www.itu.int/en/ITU-D/Statistics/Documents/facts/FactsFigures2021.pdf

Kapferer, B. (2004). Introduction: Old permutations, new formations? War, state, and global transgression. In B. Kapferer (Ed.), *State sovereignty, war: Civil violence in emerging global realities* (pp. 1–15). Berghahn.

Kaplan, A. (2000). Capacity building: Shifting the paradigms of practice. *Development in Practice, 10*(3–4), 517–526.

Kendra, J., & Wachtendorf, T. (2016). *American Dunkirk: The waterborne evacuation of Manhattan on 9/11.* Temple University Press.

Lewis, J., & Kelman, I. (2012, June). The good, the bad and the ugly: Disaster risk reduction (DRR) versus disaster risk creation (DRC). *PLoS Currents.* https://doi.org/10.1371/4f8d4eaec6af8

Lund, C. (2006). Twilight institutions: Public authority and local politics in Africa. *Development and Change, 37*(4), 685–705.

Marchezini, V. (2015a). The biopolitics of disaster: Power, discourses, and practices. *Human Organization, 74*(4), 362–371.

Marchezini, V. (2015b). Social recovery in disasters: The cultural resistance of Luizenses. In M. Companion (Ed.), *Disaster's impact on livelihood and cultural survival: Losses, opportunities, and mitigation* (pp. 293–302). CRS Press.

Martinez-Alier, J., Temper, L., Del Bene, D., & Scheidel, A. (2016). Is there a global environmental justice movement? *The Journal of Peasant Studies, 43*(3), 731–755.

Maskrey, A. (1987). *Disaster mitigation: A community based approach.* Oxfam.

Mbembe, A. (2019). *Necropolitics.* Duke University Press.

Nyberg, D. (2021, January). Corporations, politics, and democracy: Corporate political activities as political corruption. *Organization Theory, 2*(1). https://doi.org/10.1177/2631787720982618

Pickety, T. (2014). *Capital in the twenty-first century.* Harvard.

Quarantelli, E., & Dynes, R. (1977). Response to social crisis and disaster. *Annual Review of Sociology, 3,* 23–49.

Repucci, S., & Slipowitz, A. (2022). *Freedom in the world 2022: The global expansion of authoritarian rule.* https://freedomhouse.org/sites/default/files/2022-02/FIW_2022_PDF_Booklet_Digital_Final_Web.pdf

Scott, J. (1989). Everyday forms of resistance. *Copenhagen Papers in East and Southeast Asia Studies, 4,* 33–62.

Scott, J. (1990). *Domination and the arts of resistance: Hidden transcripts.* Yale University Press.

Sen, A. (1992). *Inequality reexamined.* Oxford University Press.

Shiva, V. (1999). *Biopiracy: The plunder of nature and knowledge.* South End Press.

Sousa-Santos, B. (2021). *O futuro começa agora: da pandemia à utopia.* Boitempo.

UNECE (United Nations Economic Commission for Europe). (2023). *Environmental policy. Public participation.* https://unece.org/environmental-policy-1/public-participation

Van Houtum, H. (2005). The geopolitics of borders and boundaries. *Geopolitics, 10*(4), 672–679.

Wenger, D. (1978). Community response to disaster: Functional and structural alternations. In E. Quarantelli (Ed.), *Disasters: Theory and research* (pp. 17–47). Sage Publications.

Wenger, D., & Weller, J. (1973). *Disaster subcultures: The cultural residues of community disasters. Preliminary paper no. 9, disaster research center.* University of Delaware.

White, S. (1996). Depoliticising development: The uses and abuses of participation. *Development in Practice, 6*(1), 6–15.

Wisner, B. (1988). *Power and Need in Africa: Basic human needs and development policy.* Earthscan.

Wisner, B. (1993). Disaster vulnerability: Scale, power, and daily life. *GeoJournal, 30*(2), 127–140.

Wisner, B. (1999). There are worse things than earthquakes: Hazard vulnerability and mitigation capacity in Greater Los Angeles. In K. Mitchell (Ed.), *Crucibles of hazard: Mega-cities and disasters in transition* (pp. 375–427). United Nations University Press.

Wisner, B. (2016, August 31). Vulnerability as concept, model, metric, and tool. *Oxford Research Encyclopedia of Natural Hazard Science.* Retrieved January 25, 2024, from https://oxfordre.com/naturalhazardscience/view/10.1093/acrefore/9780199389407.001.0001/acrefore-9780199389407-e-25

Wisner, B. (2020). Five years beyond Sendai – can we get beyond frameworks? *International Journal of Disaster Risk Science, 11*(1), 239–249. https://doi.org/10.1007/s13753-020-00263-0

Wisner, B., Blaikie, P., Cannon, T., & Davis, I. (2004). *At risk: Natural hazards, people's vulnerability and disasters.* Routledge.

Wisner, B., Brenes, A., & Marchezini, V. (2020). Non-governmental organizations and natural hazard governance in Latin America and the Caribbean. In *Oxford research encyclopedia of natural hazard science* (pp. 1–40). Oxford University Press.

Wisner, B., Gaillard, JC., & Kelman, I. (2012). Framing disaster: Theories and stories seeking to understand hazards, vulnerability and risk. In B. Wisner, JC. Gaillard, & I. Kelman (Eds.), *The Routledge handbook of hazards and disaster risk reduction* (pp. 18–34). Routledge.

Wisner, B., O'Keefe, P., & Westgate, K. (1977). Global systems and local disasters: The untapped power of peoples' science. *Disasters, 1*(1), 47–57.

Wisner, B., Westgate, K., & O'Keefe, P. (1976, September 9). Poverty and disaster. *New Society, 37*(727), 546–548.

Zimmerer, K. (1994). Local soil knowledge: Answering basic questions in highland Bolivia. *Journal of Soil and Water Conservation, 49*(1), 29–34.

Chapter Four
The Living World

So far, the only known planet with life is Earth – and it is saturated with it (see the Box). Humanity's physiological, sensory, and, to some extent, cognitive and cultural evolution has always occurred in the presence of and codependent with other life forms. These other living beings are necessary for human survival, even when, in some cases, they can threaten human health, well-being, lives, and livelihoods. Part of being human means interacting with other living beings: "We are one type of organism among other living things. However much our 'embodiment' . . . may be culturally or socially 'constructed' and interpreted, our existence and daily lives revolve around the care and use of bodies" (Wisner et al., 2004, p. 145).

BOX 4.1 Rare Earth, Rare Life?

It is worth pausing momentarily to consider the famous photograph of Earth taken from near the moon (see Chapter 7). That delicate, fragile-looking blue-green gemstone is teeming with life. Visible in the image are continents, clouds, and oceans (the lithosphere, the atmosphere, and the hydrosphere). All three support, nurture, transport, and exchange living things: the biosphere. Astronomers using spectroscopy are searching for planets with atmospheres containing oxygen, carbon dioxide, and methane, among other gasses, as Earth's does, on exoplanets many light-years away with orbital distances from their stars similar to Earth's that could allow the presence of liquid water. So far, out of thousands of planets studied, Earth remains the only identified sanctuary and incubator of life. The search is ongoing, and since there are likely to be trillions of exoplanets in the Milky Way

DOI: 10.4324/9781003292814-4

galaxy alone, it could turn out that life in some form will be found – possibly as different from humans as the so-called extremophiles that inhabit the Earth's deep sea volcano vents or Antarctica's Dry Valleys or perhaps as similar to humanity with similar forms, intelligence, and constructive and destructive powers.

Chapter 1 discussed several challenges to the understanding of risk and disaster. Turning to the living world in this chapter introduces another great challenge to dominant ways of seeing, feeling, and thinking. The living world demands that humanity understands itself as integral and interactive without privilege, letting go of binary constructions along the lines of people versus nature, culture versus the environment, and variations. Some cultures do not value or acknowledge dichotomous divisions that would limit biocentric understanding, although some Western cultures generally do. Nevertheless, human intellect, imagination, and emotions can grasp and communicate humanity as part of the living world.

This book's title announces the ambition to understand risk. To do so requires recognizing the flux of growth and decay, of birth and death, and of renewal and aging to which all life is subject – but, more so, recognizing that these pairs just mentioned are not meant as rigid binaries but are used for convenience. Defining "life" is subject to much debate and ambiguity (Luisi, 1998). In the context of the nonbinary approach to concepts running throughout this book, it will, therefore, seem out-of-place to divide this chapter into the micro and macro biological manifestations of the living world, as will be done in the first two parts. This binary division is yet another unfortunate convenience, whilst never denying the seamless continuum of life forms through time and across scales from the smallest prion to the possibly largest living entity, a trembling aspen "forest" called Pando comprising a single genetic individual (DeWoody et al., 2008) – or, perhaps, even the planet, according to some interpretations (Lovelock, 1972).

At the microscopic scale, organisms protect and promote human lives. Each human being contains a mass of beneficial bacteria called the microbiome weighing as much as human brains and distributed around human bodies (Reid & Greene, 2014). Other bacteria fix the nitrogen for plants valued as food. Bacteria and other microorganisms, though, can be agents of human and livestock diseases. At the macroscopic scale of the living world directly visible to us, plants and non-human animals have co-evolved with humanity as food sources and companions – and as possible reservoirs or vectors of diseases, along with other forms of threats.

Microbiology

Microbes are essential for human life and livelihoods – as well as for the function of ecosystems and the entire planetary biosphere. The human microbiome of microbes throughout the human body keeps people alive by aiding digestion, metabolism, and disease prevention among other functions. Bacteria are used to make cheese and yeast for breweries and bakeries. Fungi are a nutritious food source, often as mushrooms. Bacteria are also at the forefront of many aspects of pollution cleanup from metal contamination to oil spills, while microbial enzymes assist in medicine production. Biopesticides use microorganisms to protect crops from insects and disease, without polluting the environment with artificial chemicals. Viruses, rickettsia, fungi, algae, protozoa, nematodes, and bacteria have all been investigated for this purpose. Without microbes and other microorganisms, ecosystems could not function. The basic processing and circulation of elements, such as carbon, nitrogen, oxygen, and many more, depend on bacteria and plankton among others. Fungi decompose dead plants and animals, allowing the environment to reuse the material, most notably through creating soil.

Disease

These same classes of microorganisms can become hazards, interacting with vulnerable individuals and populations to bring about disease or even a biological disaster in the form of an outbreak, often termed an epidemic or pandemic. Besides microorganisms, there are prions and abnormal proteins that appear to lead to numerous ailments. Parasites are organisms that live off another species and cause harm in the process, such as the protozoan *Plasmodium malariae* that causes malaria.

Humans can be infected individually: a mosquito carrying a virus or parasite may bite someone and transmit the pathogen. Malaria and the viruses leading to dengue, Zika, chikungunya, and yellow fever are all vector-borne in this way. Direct human-to-human transmission of disease agents is also common, with examples including the spread of the bacteria responsible for tuberculosis, whooping cough (pertussis), and typhoid as well as the viruses responsible for flus, colds, HIV/AIDS, measles, and chickenpox (varicella).

Humans can also pick up microbial pathogens directly from the environment. The cholera bacterium and typhoid bacterium reach people through contaminated food and water. The parasitic worm causing schistosomiasis (bilharzia) lives out part of its life cycle in snails before entering humans who have contact with freshwater where the snail has shed the larvae. Infected humans may complete the cycle by defecating or urinating in water with which others have contact. Bearing in mind that the determination of thresholds is both arbitrary and culture-bound, at the scale of whole populations, the disease may be present continuously at a certain expected rate and is considered endemic. The disease may also take the form of an outbreak, perhaps leading to an epidemic or a pandemic, which are disease occurrences more than what would be expected within a population, time period, or geographic area.

Disease Control

While infectious disease is a necessary outcome of living on Earth, outbreaks can be controlled and prevented (see the Box). Vaccines and monitoring have directly and deliberately eradicated only one human disease so far: smallpox. Polio, measles, mumps, rubella, and dracunculiasis (Guinea worm disease) in people are close to eradication. In West Africa, Benin and Mali have eradicated trachoma, a bacterial infection of the eye vectored by flies that can lead to blindness due to inflamed granulation on the inner surface of the eyelid (WHO, 2023a). Yellow fever's range has shrunk hugely through a combination of surveillance, mosquito elimination, vaccines, and proof-of-vaccine requirements to travel. There is even a vaccine approved by the WHO that provides partial protection from malaria, yet the ambitious goal of eradicating malaria in Africa by 2030 faces serious obstacles (Feachem et al., 2019; WHO, 2021). The living world's dynamism and complexity are again evident in that the most lethal malaria parasite, *Plasmodium falciparum*, is developing resistance to the most common combination of treatment medications. Meanwhile, mosquitos that transmit malaria by biting humans show signs of resistance to the insecticide used on bed nets (Maharaj et al., 2013).

In terms of disease control in animals, national-scale elimination of outbreaks has been achieved across numerous diseases. Globally, in 2011, rinderpest became the first animal disease eradicated by humans; however, biological material containing the virus remains stored in various locations, so intensive vigilance is required to avoid reintroduction (Myers et al., 2018).

BOX 4.2 The COVID-19 Pandemic

The COVID-19 pandemic from 2020 to 2023 illustrates the role of vulnerability in such disasters. A virus common in some wild animals crossed to humans. An outbreak of this new human disease could have been addressed before it spread worldwide. Medical personnel in Wuhan, China (where it appears to have originated), swiftly recognized a new disease, alerted authorities about their worries, and determined detailed virological aspects (Wu et al., 2020; Zhu et al., 2020). They were met with silencing, denial, intimidation, and an attempted cover-up (Yang, 2021).

As the outbreak transformed into a severe pandemic, health system inadequacies worldwide due to long-term underinvestment and mismanagement were exposed. Many governments such as the UK had been warned of their high vulnerability to a pandemic (Public Health England, 2017) – especially due to the previous coronavirus pandemic, severe acute respiratory syndrome (SARS), from 2002 to 2004 and the H1N1 pandemic flu in 2009. The experiences and analyses did not galvanize an effort to make health systems more robust by, for instance, production and distribution of sufficient protective equipment for health care workers, measures and investments to ensure the continuity of usual health care and social services during a pandemic, and provision of accessible health care for all. Nor was the depth of vulnerability-creating actions acknowledged and tackled, from ecosystem encroachment and destruction to careful management of human-animal contact, both known to favor emerging pathogens (Garrett, 1994; Wallace, 2016; Wallace et al., 2018; Hotez, 2021). Due to such long-standing vulnerabilities, the hazard could not be addressed effectively.

While lockdowns as a measure of desperation indisputably saved numerous lives, they did not account for people's long-term vulnerabilities, especially those who were already the most vulnerable. Stress, substance use, self-harm, suicide, untreated cancer, obesity, and domestic violence were already among poorly managed epidemics, many of which worsened from the poorly managed COVID-19 pandemic. Another long-term impact of lockdowns still being monitored and assessed is delayed socialization, language development, and intellectual progress of toddlers. Small businesses which took a lifetime to build up were swiftly destroyed while many still relive the trauma of family members dying with no one holding their hand. It is representative of society, and the catastrophes humanity creates, that "social distancing" became the common phrase rather than "physical distancing", which would have encouraged remaining as socially close as possible in the absence of physical proximity.

Yet all these risks from lockdowns were known and worrisome before lockdowns were implemented. Many risks could have been averted or diminished with appropriate pandemic preparedness in the decades prior to COVID-19. By training medical staff in advance and having available enough protective equipment, other health diagnoses and treatments did not need to be slashed. Where lockdowns permitted moving abodes for specific reasons, self-harm, abuse, and domestic violence remained addressed at some level – certainly not enough nor was it enough prior to COVID-19. Encouraging outdoor exercise in many places tackled obesity and supported distanced socializing, where it is safe to be outdoors and for people who could get outside for some time each day. After all, many were so exhausted dealing full-time with kids, other families, and/or unemployment that taking care of themselves was infeasible. In some places, gym use could have been restricted with intense hygiene measures rather than banned outright. For COVID-19 and vulnerabilities, the issue was not so much the harm of lockdowns but rather the long-term lack of readiness for pandemic management. The balance between disease harm and response harm (including but not limited to lockdowns), known long before COVID-19 emerged, was far from optimum.

Environmental management and engineering have often been put to work for disease control in order to affect the hazard, as opposed to influencing vulnerability. Drainage and vector control have been used as approaches to malaria eradication for decades if not centuries, yet they have other risk consequences, notably settling ex-wetlands which possibly leads to increased flood risk (Chapter 5). The use of pesticides to control disease vectors brings with it disease risks, such as cancer. Addressing some hazards and vulnerabilities may augment others. Nevertheless, pathways circumventing such trade-offs are possible. There are ways of living on floodplains without augmenting flood risk (also treated in Chapter 5). Mosquito control can be achieved without spraying chemicals. Examples are using mosquito traps and baits, bed nets, and screens on windows and doors. Also, amphibians and fish that eat eggs, larvae, and adult mosquitoes can be useful. These are all elements of integrated vector management (WHO, 2023b). People can wear protective clothing, use repellents that do not lead to other diseases, and try to stay indoors at dawn and dusk when many bites occur. All of these measures require money and livelihoods, which many people do not have.

Vulnerability and Disease

Despite achievements in the control of disease hazards, baseline vulnerability remains. For some, this can be quite high, especially in zones of violent conflict where access to routine health care, clean water, and basic sanitation may be limited and where disruption has blocked eradication of the disease hazard. Poverty interacts strongly with disease hazards. This interaction takes the form of a vicious circle (Wisner et al., 2004, pp. 190–191). Poorer people tend to live in smaller, less well-ventilated homes with higher occupancy density and fewer hygiene and cleanliness options – all of which promote the spread of infectious disease. Resources are required for a person who is ill to take days off work, pay for medicine, or travel to a clinic. A poor, rural individual who lacks such resources might also need to continue to walk to sources of water, food, and fuel while cooking, cleaning, and caring for children and elderly relatives. Therefore, illness can have complex reverberations within household labor, livelihoods, and finances, as Barnett and Blaikie (1992) showed for HIV/AIDS. Blaikie later built upon this HIV/AIDS example in his model of resource access (Wisner et al., 2004, pp. 87–124). In addition, income poverty tends to correlate with poverty of education, poverty of information, and poverty of access to administrative and political officials (Sen, 1981). Lack of access to these human and political resources is a major source of vulnerability.

Affluence may produce vulnerability to disease hazards. With higher income, people can travel longer distances more frequently and so be infected when mixing with crowds. Travel, whether for business or pleasure, brings people who can afford it to places with microbes that they otherwise would not encounter – and provides a transport route for the microbes. Malaria cases still present themselves in local hospitals in affluent countries when people have contracted the disease elsewhere overseas or when mosquitos hitch a ride on aircraft, followed by biting and infecting people who had never left their country.

Simultaneous Pandemics – Compounding Vulnerabilities

A syndemic is a public health crisis when two or more pandemics overlap or occur simultaneously or sequentially. From 2020 to 2023, at least three infectious disease pandemics were present: COVID-19, HIV/AIDS, and cholera. Other conditions were also widely present at rates and that could easily be described as "pandemic", including obesity, malnutrition, and mental health challenges. Living with syndemics has always been "normal", even if not fully identified or admitted, but certainly clear and ever-present for the current state of the globalized "ecumene" (the known and inhabited world).

Despite the existence of health care systems, reasonable safety of food and water, acceptable standards of hygiene and sanitation, and clean homes and workplaces for many affluent populations, numerous diseases remain endemic among them. These include sexually transmitted diseases (STDs) and seasonal influenza. Despite control efforts, vector-borne diseases persist, including West Nile fever, Lyme disease, and encephalitis. In the absence of access to health care, clean food and water, appropriate hygiene and sanitation, and well-maintained homes and workplaces, many less affluent people find themselves without options to avoid such diseases daily or seasonally.

Macrobiology

Positive Roles of Flora and Fauna in Human History

Flora and fauna have long nourished, sheltered, and clothed human beings (see the Box). Wood for houses, leather for apparel, and most food have macrobiological sources. Domestication of small stock, such as goats, sheep, and chickens, supported human lives and livelihoods alongside the taming of larger animals, such as cattle, horses, and elephants, while using further creatures, such as dogs. Farming with cultivated food plants and the non-food uses of large animals began a technological spiral of innovation. Tools and techniques were developed, while the surplus food spurred the growth of settlement, trade, and eventually, large cities and widespread travel.

BOX 4.3 Domestication of Plants and Animals

When, where, how, and why humanity first started moving toward domestication of plants and animals remain subjects of extensive research and much discussion. Transition periods have been identified from hunter-gatherer to more systematic exploitation to local control to selective breeding, with overlaps, also producing debates regarding the meanings of "domestication" and "agriculture". For now, it is generally accepted that domestication had definitely started 10,000 years ago but goes back earlier, perhaps as far as 12,000 years ago. Four major independent centers of domestication appear in North America, South America, West Asia, and East Asia.

The domesticates were then transported by people as they migrated and as trade routes opened up. An intriguing question emerges. As local and regional environments changed naturally, people might have noticed that domesticated plants and animals performed better in other locations. How often did people use domesticated species to improve agriculture? Rivers change courses and volcanoes erupt, changing the land's fertility and perhaps enticing people to move their settlements.

Today, selective breeding has become a full-scale manipulation of living beings through gene editing. Genetically modified organisms (GMOs) tend to refer to those which would not occur through natural mating or offspring production. As with all technologies creating a new living world, the words of Shakespeare's Miranda are apt: "O brave new world,/That has such people in't" (Shakespeare, 1610–1611), now nongenetically modified to the "brave new world that has such more-than-humans in it".

Harms and Hazards from Flora and Fauna

Fauna and flora prey on people as food, or happen to be poisonous, toxic, or venomous. That humans can be affected in such negative ways by the surrounding macrobiota might lead to the notion that humanity exists apart from and perhaps even in opposition to plants and animals. Many phrases used – invasive species, pests, vermin, and weeds – connote negativity toward the macrobiota, making human goals and norms the top priority. In fact, a common phrase describes "human-wildlife conflicts" (Gaillard et al., 2019). Consider subsistence farmers, who encroach into crocodile or hippopotamus territory because the farmers have no other option to feed their families and lack the funds needed to erect non-harmful barriers. Then it appears and can be framed as if the animals are attacking and causing problems for humans. And who determines the division – perhaps an arbitrary binary – between "wildlife" and "domesticated animals and plants" (see the Box 4.4)?

BOX 4.4 Human-Wildlife "Conflicts", or the Dilemma of Hazards and Resources

The dominant discourse on wildlife is that wild animals found in Africa and Asia, such as the elephant, hippopotamus, rhinoceros, lion, leopard, cheetah, and buffalo, need to be protected. They constitute a symbol of "wilderness" that is currently under threat by global and local forces, such as urbanization, large-scale farming, poaching, and international trade. This imperative drives the injunction for their "protection", often articulated as being "to conserve biodiversity". This discourse resonates with the demands of tourism that earn some income for the African and Asian countries, where such wildlife is found. Tourists flock to national parks in Kenya, Tanzania, Zimbabwe, Botswana, and South Africa to see the "big five": lion, leopard, rhino, elephant, and African buffalo.

This wildlife protection discourse reflects some aspects of a Western perspective. Large mammals found in Africa and Asia symbolize a distant and pristine "other" that has long stirred the fascination of Europeans. Real dangers to these "charismatic megafauna" (e.g., Kleiman & Seidensticker, 1985) combine with concern about biodiversity loss, environmental degradation, and contemporary climate change to produce a discourse of doom. The response has been to roll out policies and programs of protection by a sophisticated dispositive that involves international treaties, national policies, non-governmental organizations, the tourism and film industries, and the media.

This discourse is being imposed as the universal and only truth, notwithstanding that some of the same animals deemed to be threatened can be seen as daily threats to local people. The majority of rural Africans and Asians do not benefit from tourism. For these people, elephants, hippos, lions, baboons, and crocodiles are a danger threatening their livelihoods and sometimes lives. Farm animals and crops are at risk. Parents may be afraid to let their children make the journey to school on their own. These hazards and disturbances from wildlife can be significant dangers, like seasonal drought or flood. Although some of these animals migrate so they do not necessarily present danger for parts of the year, others do not. Irrespective of seasonal respite or an increase of peril, awareness and worry about animal depredation are constant and unremitting.

Certainly, killing the threatening or potentially threatening animals would remove the danger. These animals are often officially protected. Hunting can be strictly regulated by governments that then earn substantial sums by selling licenses to foreign hunters to kill specific kinds of animals – even if regulations are not consistently enforced. For some local people, to kill any of the animals may lead to fines or prison, subject to connections to or bribes for officials. As a result, local people are left with small-scale, makeshift, and temporary strategies to protect themselves and their livelihoods. These include fencing, scarecrows, chili bombs and fences, and making noise with tins, flashing lights, or music. Unfortunately, clever animals may quickly overcome these challenges and manage to trespass onto people's farms, trample and eat crops, attack domestic animals, deter people from traveling on foot, especially at night, and make it risky to approach a riverside to draw water, fish, bathe, or wash clothing.

The tension between hazard and resource is evident. A possibly valid concern with declining biodiversity has led to the protection of animals that the state and outside experts consider a resource. The resource thus protected is also a hazard for ordinary people living with or alongside the animals. Moreover, the imperative to protect tends to be from a governing elite. This elite believes in protection discourse and has a material interest in regulating the "resource" by setting up protected areas and selling hunting licenses. Local voices might not be heard or are actively suppressed.

Current recommendations encourage what is termed co-existence between animals and local people, perhaps guaranteeing income from wildlife tourism and using people's local knowledges. Co-existence also involves compensation mechanisms for the losses incurred by wild animals when they destroy crops or kill livestock. Namibia constitutes a good example of this, being showcased as a game changer in managing human-wildlife contacts (Gaillard et al., 2019).

When do these harmful interactions between humans and the surrounding flora and fauna become a disaster? There are ambiguities in thinking about disaster risk that preclude an uncomplicated answer. The complications involve risk perception and the emotional, practical, and political responses to harm and hardship. A tree falling on a vehicle with one occupant or on a single pedestrian and resulting in a fatality is disastrous for the affected person, their family, and their friends. These instances are rarely included in disaster death tolls. During a storm, such as a hurricane or blizzard, individual fatalities from falling trees typically are included as well as deaths from power outages after trees fall on the lines.

Other examples of animals causing havoc for human infrastructure abound. Planes crash after hitting birds, rodents chew through cables, zebra mussels clog the water pipes of power plants, and domestic animals ignite dwelling fires. News search engines return plenty of stories of commercial flights delayed due to a cat, rat, or mouse seen on board, with concerns that the animal's interference could lead to a crash. The destruction of an aircraft would certainly be seen as a disaster. Squirrels have been known to cause electrical blackouts, with many other rodents as well as birds creating similar problems. These and other animals get into transformers and other electrical infrastructure, causing fires, arcing, broken wires, or short circuits. Rodents are implicated in food shortages, along with many other faunas. Elephants and wild pigs trample cultivated fields, while crops and stored food supplies are eaten and contaminated by rabbits, moles, and insects.

Political debates rage over other human-wildlife contacts. Do seals deplete Newfoundland fish stocks? Do kangaroos munch or trample their way through Australian farmers' fields? In some cases, animal impacts on crops are a matter of mainly economic loss, whereas loss of food or livelihood-based entitlement could be disastrous, resulting in hunger or famine. The swarms of locusts or Sudan dioch birds can decimate square kilometers of farm fields within hours. Caution is required in analyzing such hazard impacts. As Sen (1981) pointed out, there is a difference between food availability deprivation (FAD) and food entitlement deprivation (FED). He shows that the former is rare, as, most often, there are ways that political, social, or economic entitlements could make up for the immediate lack of physical availability. Likewise, the creative use of macrobiology can enhance nutrition and food availability (see the Box 4.5).

BOX 4.5 Grubs for Grub

Termites, weevils, grasshoppers, mealworms, crickets, and caterpillars can cause problems for infrastructure and damage crops. They are also packed with vitamins, protein, and minerals, meaning that they are a potential food source that many cultures around the world have tapped into throughout human history. The potential exists to make them available more widely, being a straightforward, environmentally friendly, year-round, organic source of local jobs and local food, thereby reducing the fuel used to produce and transport food. The easiest way to consume the bugs might be cooking them, through roasting, baking, frying, or boiling. Depending on how they are prepared, they might replace croutons in salad or nuts and peanuts in pastries. They can be ground up into a powder that might suit as a hot drink or cold smoothie while substituting for flour, producing bread- and cereal-type products. And it all feeds those who have nut, legume, or gluten allergies.

No food, though, could solve all needs. The possibility of allergies to insect-based food has yet to be explored fully, and it would be important to ensure proper cooking, so fuel use would need to be compared with noninsect cooking to avoid it increasing household costs. Too much of any good thing can be harmful, meaning that close monitoring of people's nutrition intake would be needed to avoid other problems arising. If a balance were achieved, then local production might become a significant contribution to sustainable nutrition and food. Of course, such food would need to be culturally acceptable (see Chapter 9), including the role of social power in shaping taste and acceptability (Counihan & Højlund, 2018).

Macro fauna and flora may harm people in other ways. The bites of some lizards, snakes, and spiders are lethal to humans and other animals. A slew of amphibians and undersea creatures are deadly; for instance, the golden poison frog in Colombia and the pufferfish that inhabits many of the world's oceans. Electric eels, orcas, and seals have been known to kill people, as have numerous large birds, reptiles, and land mammals. Ants, bees, centipedes, hornets, jellyfish, scorpions, and wasps are among the smaller creatures that may harm or kill humans and other animals, especially those who have developed an allergy to the sting or bite. Many plants are toxic to people.

Some non-human animals also transmit disease. As described earlier, biting insects, such as mosquitoes, fleas, flies, and ticks, pass microbes to us. Larger animals may play a role as reservoirs or intermediate hosts of the disease agent. For example, deer host ticks carrying Lyme disease bacteria while rats host fleas with plague bacteria. The ticks or fleas then bite people, transmitting the disease. Some birds can be a reservoir for the West Nile virus, while some birds and pigs can serve as reservoirs for influenza viruses. Coronaviruses might be harbored by bats and pangolins, whilst all mammals can host and pass on the rabies virus.

In addition to the complex and fascinating interactions of human beings with insects and microbes (Winegard, 2019; Rosen, 2008), fiction writers have imagined many disaster scenarios involving plants or animals. Examples include Wyndham's (1951) novel *The Day of the Triffids* featuring a carnivorous, mobile plant species preying on humanity after most of the population was blinded. Some emergency management agencies have used a zombie apocalypse scenario for training and education, with mixed results (Kruvand & Bryant, 2015). In the zone of speculation bridging fact and fiction, extra-terrestrial life may be found to present possibly existential hazards to humanity:

- Alien life drifting frozen through space, then thawing and becoming active on Earth.
- Returning human astronauts or a human space probe inadvertently bringing back extra-terrestrial microbes.
- Dangerous life forms unknowingly brought to Earth by friendly aliens.
- A hostile attack from outer space.

In a turnabout, the destruction of humanity by attacking Martians is averted by Earthly microbes killing the aliens in *The War of the Worlds* by Wells (1897).

Help from Animals

This chapter earlier discussed the value placed on animals by human beings throughout history. Animals were a food source, their hides provided clothing and shelter, and their bones and dried tendons could be used to make tools and ropes or used for sewing. Early human ancestors probably found a use for every part of a slaughtered animal, much as the pastoral Maasai and some other groups still do today. Dogs were used for hunting; although, as time went on, they became companions, followers, and protectors of "their" humans. That companionship continues in the form of "pets", as many other animals – including cats, birds, snakes, and fish – are pets as well. In addition, dogs with their acute sense of smell are now used to detect cancers and smuggled narcotics. Large dogs have been bred and trained to guard sheep from predators and to rescue people trapped by avalanches. Other dogs have been trained to find people buried in buildings ruined by earthquakes.

Rats have been trained to locate unexploded land mines in former conflict zones. Domesticated Asian elephants work in the timber industry in South and Southeast Asia. Cats have managed the rodent population of human settlements for millennia and, in doing so, have had a role in controlling diseases. Some breeds of geese make excellent guard animals, a fact that seems to have been known in antiquity. They are used to guard chickens from predators and thieves today. One legend, perhaps apocryphal, relates how

Tufted Roman geese sounded the alarm when the Gauls tried to capture the last part of Rome they did not control.

Ecosystems, Biomes, and Anthromes

Humans together with the living entities discussed so far in this chapter experience change along with all the other living and non-living elements around us. These wholes comprise wide-ranging collectives referred to as "ecosystems" and "biomes". An ecosystem is a network or system of interacting organisms and their non-living surroundings. The network and its environment exchange information, matter, and energy. The word biome is both a geographical and biological term describing the areal extent as well as the biotic and abiotic content of that space, typically used descriptively for one or more ecosystems across an extensive region characterized by plant and soil associations (Mucina, 2019). Echoes of the term's descriptive origins remain in some current definitions, critiques of them, and moves toward practical approaches, such as Conradi et al. (2020, p. 1296) offering, "biomes can be defined as large-scale vegetation units characterized by their growth-form spectra" in order to "link between the ranges of species and biomes".

A more recent understanding of human-nature interactions has led to developing the idea of the anthrome, a biospheric unit in space and time created by the interplay of human and non-human life together with abiotic elements. Ellis (2020, p. 5) describes the evolution from biome to anthrome:

> Anthromes (anthropogenic biomes) characterize the global patterns of life created and sustained by human societies and their use of land. These anthropogenic ecosystems now cover more than three quarters of Earth's ice-free land area, and include dense settlements, villages, croplands, rangelands, and seminatural lands; wildlands untransformed by agriculture and settlements cover the remainder.

The "anthrome" idea reinforces the assumed and artificial separation of humans from the rest of the planet by viewing human impacts and non-human impacts as different. What humans think of as a natural hazard event impacting society also has consequences for the biotic and abiotic components of ecosystems, biomes, and anthromes. Do these components cause "damage" to ecosystems and biomes or simply change them? When a windstorm fells older trees, the trees die, only to sprout new trees as the space opened admits sunlight to the forest floor while the newly dead tree provides nutrients and fertilizer for seedlings and a habitat for insects. Is it disastrous for a non-sentient being to perish as part of an ecosystem driven by time's arrow? The ongoing human devastation of species is considered by some to be a disaster and is often referred to as "the sixth mass extinction", although many others critique this framing (Cowie et al., 2022). A mass extinction affecting humans would certainly be called a disaster by any human survivors!

Bushfires started by lightning in Australia threaten ecosystems and species found nowhere else (Pickrell & Pennisi, 2020). The threat seems to be non-human. But what is a "purely environmental" process, such as the lightning-ignited fire? Vegetation zones and habitats are restricted by human activities. Where urban development or logging confines a plant or animal to a single location and then that species loses 100% of its members in a small fire or flood, is it possible to distinguish separate environmental and human causes? Human beings are part of the local biodiversity of a place. Human-formed habitats, such as farmland and cities, are part of the eco-diversity of a larger space (Barthlott et al., 1999); whilst infrastructure modifies and complements topography and hydrography, forming part of geo-diversity.

Human legal attempts to "protect" ecosystems and biomes by legal means, as discussed earlier, further bring up the issue of defining "harm". Legal rights for non-human entities, such as rivers, are being enshrined within contemporary governance regimes in Ecuador, Canada, New Zealand, and India (e.g., Boyd, 2017; Kauffman & Martin, 2018). Determining whether or not rights have been infringed readily leads to puzzles and conundrums. If an earthquake leads to a landslide that dams a river, is this a disaster infringing the river's rights? If such a disaster afflicts the river, would human action be demanded to breach the naturally-created dam? If a river already has a natural dam, perhaps forming a lake, and then an earthquake breaches this dam, have the river's rights been violated? If the natural dam was protecting a river from flooding that would change downstream ecosystems, do humans have a duty to rebuild the dam?

Such questions and conundrums reveal anthropocentrism. Human values, traits, and desires are made paramount, viewing the world with human eyes – attempting to "see" and experience the world from the perspective of a non-human living. Human metaphors are based on the five human senses. Kohn (2013) attempted to provide an "anthropology beyond the human" in his book *How Forests Think*. But he falls foul of the epistemological problem put so well by philosopher Nagel (1974) in his paper "What is it like to be a bat?" After all, these discussions are about human perceptions, expectations, and norms, even when influenced by deeply embedded connections and interactions with the environments around us.

Fire as Hazard and Resource

Fire provides an excellent illustration of the complex relationship between society and nature – and assumptions behind them as separate or as a binary. Fire is an abiotic process essential to some ecosystems, especially when it is low-intensity and periodic.

The term "wildfire" or "wildland fire" incorporates conflagration by dominant vegetation types: bushfires, forest fires, and grassfires. Other ecosystems burn, including moors, veldts, and tundra. If a flammable substance, such as oil, is on water and ignites, then marine, coastal, wetland, riverine, and lake ecosystems can be affected. Smoke and ash can settle on water as much as on land, affecting many ecosystems. The fine particulate matter in smoke contributes to respiratory illness and other health problems: 4 million people die prematurely each year due to ambient air pollution (Fuller et al., 2022), parts of which can come from fires. London's smogs, a combination of smoke from burning coal and fog, are legendary. In the 21st century, Delhi suffers annual epidemics of smoke-related disease due to farmers' burning rice stubble in their fields combined with other pollution sources, such as vehicles alongside topographic and meteorological factors and the vulnerability of low-income dwellers and the unhoused who live and sleep on the pavement (Chakrabarti et al., 2019). Brazil has faced several impacts due to fires in the Amazon rainforest (see the Box).

BOX 4.6 The Brazilian Amazon on Fire

The Amazonian region is facing devastating fires as part of deforestation. Thick smoke which had traveled thousands of kilometers from the Amazon engulfed the City of São Paulo, Brazil, on 19 August 2019. Despite reports of deforestation rates by Brazil's National Institute for Space Research (INPE, 2020), the country's president at the time, Jair Bolsonaro, dismissed them as an unreliable international campaign using "imprecise" satellite data.

At the opening ceremony of the United Nations 2019 General Assembly in New York on Tuesday, 24 September, Bolsonaro stated (Ministério das Relações Exteriores, 2019):

> my government has a solemn commitment to the preservation of the environment and of sustainable development . . . Our Amazon is larger than the entirety of Western Europe and remains practically untouched. Proof that we are one of the countries which best protect the environment.

At this time of year, the dry weather and the winds favor spontaneous and criminal fires. It is worth stressing that there are also fires started by Indigenous peoples and local populations as part of their respective cultures and form of subsistence.

It is a fallacy to say that the Amazon is the heritage of humanity and a misconception . . . to say our forest is [sic] the lungs of the world. Availing themselves of such fallacies, one or another country, instead of assisting, fell in with the press's lies and behaved disrespectfully, with a colonialist spirit. They questioned that which is most sacred to us: our sovereignty!

Smoke particles from the fires of 2019 hit the Andes before reaching the south and southeast of Brazil, resulting in even thicker smoke combined with pollution from fires in Bolivia and Paraguay. INPE (2020) reported 5,253 fires in Brazil, 1,618 in Bolivia, 1,116 in Peru, and 465 in Paraguay between 17 and 19 August. While fires are common during the summer dry season, the 2019 episodes were reportedly induced by cattle ranchers and soy farmers. They deliberately set trees ablaze to clear forests and create pasture, leading to massive burn-offs of trees in Pará state. This was part of a campaign by farmers and ranchers to draw the government's attention, claiming that cutting down trees was the only way to work (see also Raftopoulos & Morley, 2020).

Fires must have fuel, which is the vegetation, and the environmental conditions for burning, typically referred to as "fire weather" in terms of properties, such as air temperature, air humidity, and aridity of the soil and vegetation. Human and natural influences intersect as ignition sources. The main natural ignition source is lightning, with local and global trends varied and not always clear (Krause et al., 2014). People have added other ignition sources, including electric sparks and arcs, arson, vehicles backfiring, and insufficiently extinguished matches, candles, cigarettes, and campfires. Burning waste can ignite ecosystem fires. Once flames are lit, wind can drive the conflagration with embers adrift starting new fire pockets.

Fire, as with other environmental phenomena, represents a hazard, a needed ecosystem process, and a human tool simultaneously. Hunter-gatherer peoples in many places created and then managed appropriate fire regimes for their needs to support hunting, maintain trails, maintain ecosystem health, and clear fields of crop residue as well as plant disease pathogens and eggs of insects that harm crops. Ecosystem "health" in previous epochs might have meant territory kept productive with plants valued by the people managing vegetation as well as productive with the animals they hunted. Over millennia, such Indigenous peoples changed ecosystems from their "natural" state to a system more amenable to human needs.

Settlements continue to expand into burnable areas, constructing and reconstructing in locations that have burned before and that must ignite in the future (Fairbrother & Tyler, 2018). Since there is a desire to live in fire-prone areas, protective actions can be taken (Australian Building Codes Board, 2018; FEMA, 2020). Measures include planting and maintaining fire-resistant local vegetation for gardens, landscaping and slopes to direct fire away from buildings, and adhering to any locally specific fire-related legislation and governmental advice. In some places, goats have been used to reduce flammable vegetation (Mancilla-Leytón et al., 2013). For structures, examples of actions to reduce flammability include removing litter and debris from adjoining the property or in gutters while closing gaps and holes in walls and doors – all of which help to prevent embers from settling and starting a fire. Local groups can provide specialized advice to assist with local conditions (see the Box 4.7).

BOX 4.7 Wildfire Partners

An organization called Wildfire Partners is funded by local and state governments, helping residents in Colorado, USA, to improve the fire resistance of their properties. Wildfire Partners provide risk assessments, advice for reducing risks, progress checks on implementing the advice, and sometimes monetary support for making changes. These actions are for households and families, with Wildfire Partners encouraging community approaches as well. Participants are supported for bringing neighbors together at local meetings, to consider what could be done for fire resistance along property boundaries with both owners, and to improve neighborhood roads. The ripple effect is used: one successful homeowner can spread the message to neighbors and entice them to act in emulation.

Results were seen during the July 2016 Cold Springs Fire in the mountains above Boulder. In the burned area, eight houses had participated in the Wildfire Partners program and all of them survived, remaining habitable afterward. Other houses in the area which had not taken fire resistance measures were destroyed and had to be rebuilt.

As should be expected, no guarantees are ever available that a building will avoid becoming ash. It is possible to remain behind during a fire warning in order to assist with defending a property, for instance, by hosing down the roof and walls. This choice increases the chances of properties surviving and also increases the chances of the property protector dying, leading to discussions of whether to stay and defend or to evacuate (Haynes et al., 2010).

Much of fire management, especially extinguishing them, relies on reducing heat. Yet air temperatures around the world are getting hotter due to human-caused climate change. Vegetation will be set alight more readily and then will burn more quickly and with higher intensity, while stronger winds will disperse embers and smoke farther afield. The changed environmental conditions will impede ecosystem recovery after a fire because the plants must establish and grow in a different environment to the one in which they evolved. Higher temperatures are bringing severe fires to places that have rarely experienced before, notably in the higher latitudes where tundra in Siberia and forests in Kalaallit Nunaat (Greenland) have been experiencing unusual blazes that are highly damaging to these ecosystems. When steep slopes are denuded by fire, rain may bring landslides as a secondary hazard, and the natural regeneration of vegetation cover will be delayed.

More Than the Living World?

As shown in this chapter, the living world is dynamic and hugely intertwined with non-living aspects and human social worlds. The resulting interplay among human life, non-human life, and abiotic environments is part of understanding disaster risks, hazards, and vulnerabilities. At the micro-scale, microbes pervade human bodies and living spaces. They are essential for maintaining human and ecosystem health and are key elements in many livelihood activities. Microbes can nonetheless harm humans through various pathways. This chapter also revealed the interdependence of humans on many large animals and plants, as well abiotic components of the world, including their possible harm and the pathways to this harm.

Biases and ways of seeing emerge regarding presumed binaries, such as life versus non-living. Some disease-causing agents, notably viruses and prions, straddle the realms of the biotic and the abiotic. Problems of classification remind of the need to seek multicultural insight and balance in understanding the human and more-than-human (see Chapter 8). Speculate how human ancestors viewed nature, with their cultures and actions as part of their ecosystems and biomes! Some humans still struggle to grasp and express the more-than-human being intimately entwined with human life and deeply anchored in space and time. Memory, imagination, and belief are all part of the interconnected living and non-living world, affecting what is perceived as a hazard or a resource. Human senses of place, time, life, and living remain elements of identities and worldviews. The sublime, the useful, the harmful, and the horrific are all possible in the living/non-living world shared around planet Earth.

References Cited

Australian Building Codes Board. (2018). *AS 3959:2018, Construction of buildings in bushfire-prone areas.* Australian Building Codes Board. https://www.standardsau.com/preview/AS%203959-2018.pdf

Barnett, A., & Blaikie, P. (1992). *AIDS in Africa: Its present and future impact.* Belhaven.

Barthlott, W., Biedinger, N., Braun, G., Feig, F., Kier, G., & Mutke, J. (1999). Terminological and methodological aspects of the mapping and analysis of the global biodiversity. *Acta Botanica Fennica, 162,* 103–110.

Boyd, D. R. (2017). *The rights of nature: A legal revolution that could save the world.* ECW Press.

Chakrabarti, S., Khan, M., Kishore, A., Roy, D., & Scott, S. (2019). Risk of acute respiratory infection from crop burning in India: Estimating disease burden and economic welfare from satellite and national health survey data for 250 000 persons. *International Journal of Epidemiology, 48*(4), 1113–1124. https://doi.org/10.1093/ije/dyz022

Conradi, T., Slingsby, J. A., Midgley, G. F., Nottebrock, H., Schweiger, A. H., & Higgins, S. I. (2020). An operational definition of the biome for global change research. *New Phytologist, 227*(5), 1294–1306. https://doi.org/10.1111/nph.16580

Counihan, C., & Højlund, S. (2018). Making food public: An ethnographic approach. In C. Counihan & S. Højlund (Eds.), *Making food public: Ethnographies of food and the senses* (pp. 1–10). Bloomsbury.

Cowie, R. H., Bouchet, P., & Fontaine, B. (2022). The sixth mass extinction: Fact, fiction or speculation? *Biological Reviews, 97*(2), 640–663. https://doi.org/10.1111/brv.12816

DeWoody, J., Rowe, C. A., Hipkins, V. D., & Mock, K. E. (2008). "Pando" lives: Molecular genetic evidence of a giant aspen clone in Central Utah. *Western North American Naturalist, 68*(4), 493–497. https://doi.org/10.3398/1527-0904-68.4.493

Ellis, E. (2020). Anthromes. In M. Goldstein & D. DellaSala (Eds.), *Encyclopedia of the world's biomes: Vol. 5. Sustainability in concept and practice* (pp. 5–11). Elsevier.

Fairbrother, P., & Tyler, M. (Eds.). (2018). *Wildfire and power: Policy and practice.* Routledge.

Feachem, R. G. A., Chen, I., Akbari, O., Bertozzi-Villa, A., Bhatt, S., Binka, F., Boni, M. F., Buckee, C., Dieleman, J., Dondorp, A., Eapen, A., Feachem, N. S., Filler, S., Gething, P., Gosling, R.,

Haakenstad, A., Harvard, K., Hatefi, A., Jamison, D., . . . Mpanju-Shumbusho, W. (2019). Malaria eradication within a generation: Ambitious, achievable, and necessary. *The Lancet, 394*(10203), 1056–1112. https://doi.org/10.1016/S0140-6736(19)31139-0

FEMA. (2020). *Creating a community wildfire protection plan.* FEMA (Federal Emergency Management Agency). https://www.usfa.fema.gov/downloads/pdf/publications/creating_a_cwpp.pdf

Fuller, R., Landrigan, P. J., Balakrishnan, K., Bathan, G., Bose-O'Reilly, S., Brauer, M., Caravanos, J., Chiles, T., Cohen, A., Corra, L., Cropper, M., Ferraro, G., Hanna, J., Hanrahan, D., Hu, H., Hunter, D., Janata, G., Kupka, R., Lanphear, B., . . . Yan, C. (2022). Pollution and health: A progress update. *The Lancet: Planetary Health, 6*(6), E535–E547. https://doi.org/10.1016/S2542-5196(22)00090-0

Gaillard, JC., van Niekerk, D., Shoroma, L. B., Coetzee, C., & Amirapu, T. (2019). Wildlife hazards and disaster risk reduction. *International Journal of Disaster Risk Reduction, 33*, 55–63. https://doi.org/10.1016/j.ijdrr.2018.09.009

Garrett, L. (1994). *The coming plague: Newly emerging diseases in a world out of balance.* Farrar, Straus and Girouz.

Haynes, K., Handmer, J., McAneney, J., Tibbits, A., & Coates, L. (2010). Australian bushfire fatalities 1900–2008: Exploring trends in relation to the 'prepare, stay and defend or leave early' policy. *Environmental Science & Policy, 13*(3), 185–194. https://doi.org/10.1016/j.envsci.2010.03.002

Hotez, P. (2021). *Preventing the next pandemic.* Johns Hopkins University Press.

INPE. (2020). *Programa Queimadas.* https://queimadas.dgi.inpe.br/queimadas/portal

Kauffman, C. M., & Martin, P. L. (2018). Constructing rights of nature norms in the US, Ecuador, and New Zealand. *Global Environmental Politics, 18*(4), 43–62. https://doi.org/10.1162/glep_a_00481

Kleiman, D. G., & Seidensticker, J. (1985). Pandas in the wild: The giant pandas of wolong. George B. Schaller, Hu Jinchu, Pan Wenshi, and Zhu Jing. University of Chicago Press, Chicago, 1985. xx, 298 pp., illus. $25. *Science, 275*, 875–876. https://doi.org/10.1126/science.228.4701.875

Kohn, E. (2013). *How forests think: Toward an anthropology beyond the human.* University of California Press.

Krause, A., Kloster, S., Wilkenskjeld, S., & Paeth, H. (2014). The sensitivity of global wildfires to simulated past, present, and future lightning frequency. *Journal of Geophysical Research – Biogeosciences, 119*(3), 312–322. https://doi.org/10.1002/2013JG002502

Kruvand, M., & Bryant, F. B. (2015). Zombie apocalypse: Can the undead teach the living how to survive an emergency? *Public Health Reports, 130*(6), 655–663. https://doi.org/10.1177/003335491513000615

Lovelock, J. (1972). Gaia as seen through the atmosphere. *Atmospheric Environment, 6*(8), 579–580.

Luisi, P. L. (1998). About various definitions of life. *Origins of Life and Evolution of Biospheres, 28*, 613–622. https://doi.org/10.1023/A:1006517315105

Maharaj, R., Raman, J., Morris, N., Moonasar, D., Durrheim, D. N., Seocharan, I., Kruger, P., Shandukani, B., & Kleinschmidt, I. (2013). Epidemiology of Malaria in South Africa: From control to elimination. *The South African Medical Journal, 103*(10), 779–783. https://doi.org/10.7196/SAMJ.7441

Mancilla-Leytón, J. M., Pino Mejías, R., & Martín Vicente, A. (2013). Do goats preserve the forest? Evaluating the effects of grazing goats on combustible Mediterranean scrub. *Applied Vegetation Science, 16*(1), 63–73. https://doi.org/10.1111/j.1654-109X.2012.01214.x

Ministério das Relações Exteriores. (2019, September 24). *Speech by Brazil's President Jair Bolsonaro at the opening of the 74th United Nations general assembly.* https://www.gov.br/mre/en/content-centers/speeches-articles-and-interviews/president-of-the-federative-republic-of-brazil/speeches/speech-by-brazil-s-president-jair-bolsonaro-at-the-opening-of-the-74th-united-nations-general-assembly-new-york-september-24-2019-photo-alan-santos-pr

Mucina, L. (2019). Biome: Evolution of a crucial ecological and biogeographical concept. *New Phytologist, 222,* 97–114. https://doi.org/10.1111/nph.15609

Myers, L., Metwally, S., Marrana, M., Stoffel, C., Ismayilova, G., & Brand, T. (2018). *Global rinderpest action plan: Post-eradication.* The Food and Agriculture Organization of the United Nations and the World Organisation for Animal Health. https://www.fao.org/3/CA1965EN/ca1965en.pdf?eloutlink=imf2fao

Nagel, T. (1974). What is it like to be a bat? *The Philosophical Review, 83*(4), 435–450. http://www.jstor.org/stable/2183914

Pickrell, J., & Pennisi, E. (2020). Record U.S. and Australian fires raise fears for many species, *Science, 370*(6512), 18–19. https://doi.org/10.1126/science.370.6512.18

Public Health England. (2017). *Exercise cygnus report. Tier one command post exercise pandemic influenza. 18 to 20 October 2016.* Public Health England. https://www.gov.uk/government/publications/uk-pandemic-preparedness/exercise-cygnus-report-accessible-report

Raftopoulos, M., & Morley, J. (2020). Ecocide in the Amazon: The contested politics of environmental rights in Brazil. *The International Journal of Human Rights, 24*(10), 1616–1641. https://doi.org/10.1080/13642987.2020.1746648

Reid, A., & Greene, S. (2014). *FAQ: Human microbiome.* American Society for Microbiology. https://asm.org/Reports/FAQ-Human-Microbiome

Rosen, W. (2008). *Justinian's flea: The first great plague and the end of the Roman empire.* Penguin.

Sen, A. (1981). *Poverty and famines: An essay on entitlements and deprivation.* Oxford University Press.

Shakespeare, W. (1610–1611). *The tempest.* Performed.

Wallace, R. (2016). *Big farms make big flu.* Monthly Review Press.

Wallace, R., Chaves, L., Bergmann, L., Ayres, C., Hogerwerf, L., Kock, R., & Wallace, R. G. (2018). *Clear-cutting disease control: Capital-led deforestation, public health austerity, and vector-borne infection.* Springer.

Wells, H. G. (1897). *The war of the worlds.* William Heinemann.

WHO (World Health Organization). (2021). *Global technical strategy for malaria 2016–2030, 2021 update.* https://www.who.int/publications/i/item/9789240031357

WHO (World Health Organization). (2023a, May 16). *WHO congratulates Benin and Mali for eliminating trachoma as a public health problem.* https://www.who.int/news/item/16-05-2023-who-congratulates-benin-and-mali-for-eliminating-trachoma-as-a-public-health-problem

WHO (World Health Organization). (2023b). Control of neglected tropical diseases. *Vector Control.* https://www.who.int/teams/control-of-neglected-tropical-diseases/interventions/strategies/vector-control

Winegard, T. (2019). *The Mosquito: The human history of our deadliest predator.* Dutton.

Wisner, B., Blaikie, P., Cannon, T., & Davis, I. (2004). *At risk: Natural hazards people's vulnerability and disasters* (2nd ed.). Routledge.

Wu, F., Zhao, S., Yu, B., Chen, Y.-M., Wang, W., Song, Z.-G., Hu, Y., Tao, Z.-W., Tian, J.-H., Pei, Y.-Y., Yuan, M.-L., Zhang, Y.-L., Dai, F.-H., Liu, Y., Wang, Q.-M., Zheng, J.-J., Xu, L., Holmes, E. C., & Zhang, Y.-Z. (2020). A new coronavirus associated with human respiratory disease in China. *Nature, 579,* 265–269. https://doi.org/10.1038/s41586-020-2008-3

Wyndham, J. (1951). *The day of the Triffids.* Michael Joseph.

Yang, X. (2021). Domestic contestation, international backlash, and authoritarian resilience: How did the chinese party-state weather the COVID-19 crisis? *Journal of Contemporary China, 30*(132), 915–929. https://doi.org/10.1080/10670564.2021.1893557

Zhu, N., Zhang, D., Wang, W., Li, X., Yang, B., Song, J., Zhao, X., Huang, B., Shi, W., Lu, R., Niu, P., Zhan, F., Ma, X., Wang, D., Xu, W., Wu, G., Gao, G. F., Tan, W., & China Novel Coronavirus Investigating and Research Team. (2020). A novel coronavirus from patients with pneumonia in China, 2019. *The New England Journal of Medicine, 382,* 727–733. https://doi.org/10.1056/NEJMoa2001017

Chapter Five
Water, Weather, and Climate

Water and air are essential for human life and livelihoods, as they are for other living beings (see Chapter 4). Religious and cultural symbolism and ritual around the world and throughout history have often involved or invoked water and air. Arts depict water and air in the form of wind as friend and foe, as life-giver and life-taker. Even though the vast majority of the Earth's surface is covered by water of some form, most comprises oceans or seas, making it too salty for human beings to consume directly. The next largest fraction is frozen, making it difficult for many uses. Meanwhile, copious amounts of groundwater require extensive effort to access.

Freshwater is defined as having fewer than 500 ppm of dissolved salts (Albert et al., 2021) and covers just 2.3% of the planet's surface when considering lakes, reservoirs, and rivers. Freshwater wetlands cover at least 3.8% more (Lehner & Dölla, 2004). Humanity attempts to control much of this freshwater. Dudgeon (2019) summarizes that, globally, large reservoirs capture five times the volume of water in rivers, and almost half of the world's river volume undergoes significant flow regulation. All this water on Earth, the hydrosphere, interacts with the air surrounding the planet, the atmosphere, involving liquid and frozen water as well as gaseous water in the form of water vapor. Together, they form the air and water resources giving and sustaining life, while being the basis of numerous hazards.

Water

Water in Excess

As vital as water is to human life, too much can cause difficulties when lacking preparedness, planning, damage mitigation, and vulnerability reduction. In English, terms used to

DOI: 10.4324/9781003292814-5

describe excess liquid water on land include "flood" and "inundation". They arise from rain falling and running off, melting snow and ice, and wind and low atmospheric pressure pushing lake and ocean water onto coastlines (storm surge). The question of who actually defines "excess" and what assumptions are used to set the threshold beyond which there is an "excess" brings to mind the importance of different cultural definitions, as discussed in Chapter 3. Inundations may be followed by biological hazards (see Chapter 4). Snakebites and rodent bites during floods are common. Urban flooding in Australia brings funnel web spiders into contact with humans and the same for fire ants in Texas. Microbiological consequences of ingestion of flood water are common, including cholera, leptospirosis, and digestive illnesses from *E. Coli* bacteria. New diseases are rare following floods, although already endemic diseases, such as malaria and other mosquito-borne diseases, can reemerge as outbreaks or expand as epidemics due to increased mosquito breeding in flood waters if adequate countermeasures are not taken.

Living with Water Excesses

Living with water excesses can be attempted with structures that separate people and water, such as walls, dams, dikes, levees, and reservoirs. In the absence of actively reducing vulnerability, relying on such structural approaches often increases rather than decreases disaster risk because people remain or settle in flood-prone areas with misplaced confidence in the structures that they assume will protect them. This phenomenon is called "risk transference" (Etkin, 1999). Risk appears to be lowered immediately due to an intervention, yet it increases in the long-term; hence, risk is transferred from the present to the future.

When structural measures separate people and water, inhabitants near coastlines, waterways, and wetlands become inured to the absence of water excesses. Few floods occur, so flood risk reduction actions tend to lapse without support for continual disaster risk reduction activities. Decreased awareness, fewer risk mitigation actions, and reduced readiness for timely response augment vulnerabilities to excess water, even if the water-related dangers decrease overall. Thus, flood risk may increase as flood hazard diminishes.

In many instances, people do not even have much of a choice: they cannot afford to live elsewhere; they have no say in the structures being built; and/or they have been denied the knowledge, resources, and opportunities to tackle the factors increasing their own flood risk. Then, when a large flood rips past or through the structures separating people and water, vulnerability and risk are so high that a catastrophe results (Etkin, 1999; Tobin,

1995). However, risk transference does not preclude the benefits of structural measures. London, Singapore (as described in the Box 5.1), and Rotterdam have gained significant land and become world ports partly supported by huge sea barrages and significant river walls. In some places, reservoirs store water for times of need and provide recreational sites. These benefits should not be denied nor should the long-term costs of risk transference be underemphasized.

BOX 5.1 Singapore's Marina Barrage

The city of Singapore in the island country of the same name sits at the confluence of five rivers producing a bay that meets the sea. Or it would meet the sea apart from the 350 m long Marina Barrage separating Singapore and the sea. The barrier serves many purposes:

- It reduces the saline content of the bay, helping to supply the country with freshwater.
- It prevents many parts of the city from being flooded during high tides and storm surges.
- It prevents many parts of the city from being flooded during high rainfall by draining the water.
- It provides recreational areas through walking, boating, and relaxing.

The Marina Barrage has also encouraged significant building development in floodable areas that used to be wetlands and floodplains.

A structure-focused way of living with excess water is dubbed "flood control", "flood protection", and "flood defense" – redolent of the modern technological episteme (see Chapter 3) – although only a finite level of control, protection, and defense can ever be provided. A more realistic term would be "flood alteration" or, with such neutrality eschewed by Criss and Shock (2001), "flood enhancement through flood control" and by Tobin (1995) as "Levee Love Affair". Similarly, referring to "flood proofing" can instill a sense of invulnerability, yet it is only truly feasible in the complete absence of water. More balanced terminology includes "flood damage reduction", "flood risk management", and "flood resistance measures". Despite drawbacks, the suggestion is not that structural approaches are always inappropriate and undesirable. Instead, disaster risk reduction means advocating for honesty regarding the advantages and disadvantages of each choice while seeking a balance of actions together. It can be achieved through discussions among outside experts and inside experts (local people) built on trust and mutual respect (Wisner, 1995).

"Space for water" (Szöllösi-Nagy & Zevenbergen, 2005) has been accepted by people living on small river islands (*chars*) in Bangladesh and for the contemporary development of cities, such as Toronto and Amsterdam. In Bangladesh, the char dwellers are flexible and mobile. As the river erodes or accretes land, the people shift with the sediment or accept getting wet, thus living with local vagaries rather than being impeded by them. Toronto stopped development in several ravines after Hurricane Hazel killed over 80 people in 1954 and turned these floodplains into areas for nature and recreation (Doberstein et al., 2019). As farms that had earlier absorbed rainfall were transformed into shopping malls, office centers, and dwellings, runoff intensified and the volume of floodwater coursing through the ravines increased. Risk did not rise excessively due to zoning control in the ravines that kept vulnerability low.

Humans can be inventive in the way they live in floodplains. The Netherlands has pioneered floating homes which rise and fall with the water level. Although this solution might be less appropriate for bodies of water subject to waves and swift currents, Bolonkin (2011) describes floating cities. They could be fixed in place with anchors or free to drift around the world. Fiction has long depicted this way of living with water, both in novels such as Baxter's (2008) *Flood* and movies such as "Waterworld" (Herzberg et al., 1995).

Some cultures are currently more amphibious. The Moken people are nomadic, living on watercraft throughout the archipelagos around the southwest Burmese-Thai border (Dancause et al., 2009). Excess water would be hard for them to define because they routinely change their navigation patterns as needed without it being either worrisome or a boon. Knowledge of their water-based environment limited casualties on 26 December 2004, when they recognized and responded to the impending signs of a tsunami (Arunotai, 2008) that killed approximately 250,000 people around the Indian Ocean. Underwater living is another option presented by science (Kaji-o'grady & Raisbeck, 2005; Riffat et al., 2016) and in fiction (e.g., the city Otoh Gunga on the planet Naboo in "Star Wars: Episode I – The Phantom Menace" by Lucas, 1999).

Water in Deficit

Water deficits are typically referred to as "droughts", although as with water excesses, a "deficit" is a subjective and culturally dependent label. Many locations have perpetually low precipitation with names ascribed, including drylands, arid lands, and deserts, which refer to about one-third of the Earth's land (Goudie, 2006). Measuring the area of the Earth's drylands is not straightforward. Much of Antarctica's land and seas are covered by ice, but the precipitation rate is so low that it is termed the driest continent and the world's largest desert.

In relation to human needs and wants, water deficits may be specified in a variety of ways. "Drought" may be parsed into levels of aridity that affect the average or normally expected conditions for agricultural production ("agronomic drought" or "agricultural drought") as opposed to a statistically significant deviation of precipitation and/or snow and ice melt from the expected mean ("meteorological drought"). Where the amount of water available dips below long-term averages – perhaps due to lower precipitation, less snow and ice melt, or sea levels lowering – continuing the established rates of water use makes it seem as if a water deficit is present. Conversely, water use might increase while the amount of available water remains the same or increases, making it seem as if a water deficit exists. Extensive amounts of the world's groundwater have been used up, with extraction rates exceeding recharge rates, so places relying on aquifers appear to be affected by drought when the problem is really water overuse. Droughts are complicated, with different causes and categories leading to the same apparent outcome (Glantz & Katz, 1977; Wilhite & Glantz, 1985).

Consider droughts on opposite sides of the Pacific Ocean in 2015. Both California and Taiwan lacked enough water to match their rates of usage. In California, industrial and agricultural use has been sucking up vast amounts of surface and groundwater, while many homes require continual irrigation for their lawns and gardens due to landscaping with grasses and trees imported from wetter climates. Los Angeles, in particular, grew on the assumption that water can be pumped in from elsewhere, such as Owens Lake, located 300 km north of the city. As this lake became depleted, its local environment and the Indigenous people living nearby have been detrimentally impacted (Bertenthal, 2021). Taiwan's need for freshwater rationing during the 2015 drought was partly caused by an old and leaky water delivery system and reservoirs that were blocked by silt and sediment buildup. Precipitation variations bring water deficits to the forefront, yet the baseline cause often remains human decisions about managing the water available.

Water deficits or apparent water deficits interact with other human decisions, notably land use management, to produce drought-type conditions. People in Africa's Sahel region have long managed nomadism and farming during wide precipitation variations. Then, from around 1968–1974, tens of thousands died in a drought while others survived, mainly due to external humanitarian assistance. While rainfall had changed during this time period, analyses pointed to social changes and the lingering legacy of colonialism as the real causes of the disaster, irrespective of the water deficit. Increases in the number of people managing the same land increased the number of livestock, while the imposition of new international borders, forced sedentarization, and state policies that encouraged export rather than food crops combined to undermine agro-pastoralism and nomadism (Comité d'Information Sahel, 1975; Copans, 1975; Glantz, 1976).

Droughts may degrade soil quality with subsequent impacts affecting numerous ecosystems and livelihoods. Soil erosion changes the sedimentation load of rivers, in turn affecting coastal ecosystems when excess sediment is dumped at the river's mouth. If dryness kills vegetation, wind can scour away the topsoil, reducing the land's productivity even further and producing dust and sandstorms far away.

Living with Water Deficits

Fleuret (1986) explains that "Drought is a subjective phenomenon" (p. 224) because it depends on whether or not people are ready and able to deal with changes to water availability. Examples exist of living with water deficits to the extent that they are not really "deficits" but simply accepted environmental variations. From the Sahel and sub-Saharan Africa, Fleuret (1986) and Mortimore (2010) explain how, at least until externally imposed changes interfered with and undermined the peoples' cropping patterns, techniques included crop diversity, interplanting, and flexibility of livelihood sources to help people get through water deficits. People's foresight and preparation are evident in the common production of so-called "famine foods", such as mandioc (cassava) and yams, plants that produce food during times of water-deficit-related need. Those foods preserved and stored in advance are then available when others fall short. Wild vegetation, nuts, and berries are used as supplemental food sources. Agricultural and livestock diversity further supports some livelihoods and food for a while, no matter what the rainfall.

Another strategy used by small farmers is the production of what is called "normal surplus" (Allan, 1977). Field size and cropping patterns are chosen so that a surplus is produced in a year with average rainfall, which can be stored or sold. In a year of moderate deviation below the average rainfall, sufficient food for household needs will still be produced. Only in excessively dry years would production fail to meet needs.

Fleuret (1986) and Mortimore (2010) suggest that localized droughts were dealt with through systems of food sharing and exchange, helping out neighbors when they were in trouble so that the neighbors could reciprocate at other times. Additionally, people would farm numerous smaller plots of land across a wide area rather than a single larger farm. If one plot failed or local conditions were difficult, then people would migrate to other plots in different climatic zones, ensuring that some food would always be available somewhere. Access to these dispersed plots could be governed by kinship ties, market relations such as rental or ownership, and conventional tribal usufruct (Smucker & Wisner, 2008). A so-called "drought" may become part of the regular rhythm of life rather than a concern, a resource deficit, or a disaster – as long as techniques were sufficient to deal with the changed conditions.

Even when some water is available, difficulties may emerge due to forced resettlement, land management legislation, consolidation of land ownership, and constraints on temporary movements. Traditional knowledge may be lost, opportunities to implement it may disappear, and food habits might change in favor of imported, non-local foods, thus increasing dependence on faraway supply chains and price fluctuations.

When water is lacking, families must spend more time fetching water. Calories are burned to go farther afield for water, meaning that more food must be consumed, while children assisting with water collection have less time for studying or simply drop out of school. In areas with violent conflict, trekking along paths could expose water collectors to assault. The need to prioritize drinking water for people and animals can lead to declining water use for hygiene and safe cooking. Household-sized water tanks capturing water during the rainy season can sustain families through an extended dry season, but they are heavy to transport and with appropriate management might support insect breeding. An example of overcoming the challenges is locally constructed tanks made from chicken wire coated with a mixture of earth and cement and with adequate sealing.

Nomadic lifestyles as a response to arid conditions are not a panacea. Some people in the southern African country of Malawi follow shifting rains; however, this strategy has not been sufficient for many who have ended up pursuing precarious livelihoods (Lewin et al., 2012). Despite an elaborate system of livestock management, water storage, and livelihood diversity, the Indigenous Santal people in northwest Bangladesh struggle during droughts (Ahmed et al., 2019). Some of the world's river systems (e.g., Danube, Rio Grande, La Plata, Yangtze, Saleen-Nu, Mekong, and Ganges) are facing a range of challenges, including dam construction, pollution, overfishing, and biodiversity decline (CBD, 2003). Many aquifers are being drained at a rate above their recharge rate (Konikow & Kendy, 2005; see the Box 5.2). As always, dealing with environmental variations means that the changes can be hazardous or opportunities – and, often, both depending on who is being affected.

BOX 5.2 Iran as "Dryland"

Around 90% of Iran is arid. Where water is available, such as around Tehran, it is generally upwelling due to earthquake faults bringing with it significant seismic dangers. Water has made the location attractive for living despite the high seismic risk (Jackson, 2001), so Tehran has expanded, becoming a focus for the surrounding area (and the country's capital city) – and has taxed the water supply, leading to a long-term water deficit. Traditionally, Iran supported urban areas' water supply through "qanats", underground tunnels transporting water from the mountains. Their construction minimized losses from evaporation and contributed to cooling traditional dwellings to help people live with hot temperatures. As population numbers rose alongside water use per capita, damming of and drawdown from mountain water sources has limited the effectiveness of traditional qanats. Freshwater supply and indoor temperature control have been inhibited, while much infrastructure around Tehran and the rest of Iran remains inadequate to withstand the expected tremors.

Weather

Weather is defined as the state of the atmosphere. It is direct experience of the air, such as temperature, humidity, wind speed and direction, and precipitation. Weather certainly affects how water is distributed as well as many other phenomena.

Weather Phenomena

High winds can lead to extensive damage, ripping off roofs, blowing in windows and doors, toppling vehicles on roads, and making it dangerous to walk. Metal sheeting, a cheap and convenient roof material often used in tropical locations, can swirl through the air. Wind can bring dust storms, erode soils, obscure visibility, damage equipment and infrastructure, and inhibit breathing. Ironically, some of this dust is beneficial in supporting ecosystems when Sahara sand crosses the Atlantic to fertilize the Amazon River basin and Caribbean islands. When combined with precipitation, wind can cause brutal driving rain, augmenting dampness in building envelopes and producing damage from the physical impact on infrastructure, vehicles, people, animals, and vegetation.

Wind also produces rotating storms, including cyclones and tornadoes. For aircraft, the name or source of convection or rotation is less important than avoiding it, so airports have strict protocols regarding aircraft near wind shear, thunderstorms, and possible vortices (FAA, 2015). Within storms, water may appear in solid forms: freezing rain, graupel, hail, ice, rime, sleet, slush, and snow. The physical impact of solid water falling from the sky can damage crops, infrastructure, and vehicles – including aircraft – while injuring or killing people and animals. Some of the solid water stays on roads and sidewalks, making transportation treacherous and may accumulate on power lines and lead to power outages.

In January 1998, parts of eastern North America were coated by an ice storm, knocking out power to large swathes for days and some areas for weeks. To survive without power, many people used generators, leading to numerous fatalities from carbon monoxide poisoning. These situations are not unusual (see the Box 5.3). Another danger with winter is cold temperatures, so that anyone injured by slipping on ice or in a vehicle crash might quickly succumb to hypothermia without prompt assistance.

Wind augments the cold through the "wind chill factor", making it feel much colder than the air temperature. For instance, when a wind of 40 km/h is reached, making umbrellas hard to use, 0°C feels like -7°C and –15°C is like -27°C, verging on high frostbite risk.

BOX 5.3 Canada's Ice Storms

Canada has had numerous ice storms with major power outages.

- March 1958: St. John's, Newfoundland, experienced freezing rain for 43 hours straight.
- February 1961: Parts of Montréal, Québec, lost power for a week due to an ice storm.
- January 1968: Power failures followed three days of snow and ice across areas of southern Ontario.
- April 1984: Power went out for several days in St. John's, Newfoundland, following an ice storm.
- December 1986: An ice storm knocked out electricity to one-quarter of homes in Ottawa, Ontario.
- January 1998: An ice storm left millions of people without power, some for several weeks.

Source: Etkin & Maarouf (1995)

Repeated cycles of freezing and thawing produce other risks. Water infiltrates any gaps or cracks in infrastructure, so if the temperature then drops so that the water freezes, it expands and puts pressure on the structure. Several freeze-thaw cycles can end up breaking small parts of concrete or stone, leading to significant damage.

Cold weather affects people and households differently. In cold conditions, a poor household may have to choose between paying for heat or satisfying other basic needs, thus producing the conundrum, of "to heat or eat" (Burlinson et al., 2022). Heating with wood can lead to indoor air pollution and can augment fire risk. Differential vulnerability also manifests biologically. Well-nourished and healthy people are able to withstand cold better than those who are malnourished and in poor health. Cold, damp interior conditions may exacerbate some illnesses or delay recovery. People with adequate information and access to government or non-governmental services may be able to obtain assistance in paying for their home heat; however, such assistance is also determined by government policy making and budgeting. Some people such as detainees, migrants labeled "illegal", and those experiencing homelessness may not be in a position to claim such entitlements. The impact of winter weather is a function not only of the cold-related hazards but also of vulnerability.

Heat kills by causing people to overheat and dehydrate. As humidity increases, air temperature feels higher and perspiration evaporates less, so death is possible at lower air temperatures. Direct burns from the sun may be severe. The length of time someone experiences a high heat-humidity combination contributes to the level of danger. Even broiling during the hottest hours of the day might not kill if one is not exerting oneself and then gains a level of relief as the evening cools. If the heat-humidity combination (humidex) does not fall at night, then mortality can shoot up without artificial cooling. Not all artificial cooling, though, is helpful. When the humidex hits a point when fans simply blow hot and humid air onto people, it accelerates dehydration and kills more quickly. Air-conditioning dumps interior heat from structures into the environment nearby (Davis, 2006), increasing the urban heat bubble. People experiencing homelessness who live on the streets can be severely harmed, including by automobile air conditioners contributing to the waste heat expelled into the surroundings. Once again, vulnerability and socially constructed risk are at play for disasters – not simply the weather.

When it comes to the notion of a "heat wave", a precise definition is elusive. The diurnal humidex over successive days tends to be the best indication of the level of danger – keeping in mind that people's physiology, exposure to the sun, level of exercise, and access to cool living and working space factor in. IPCC (2021–2022) is understandably noncommittal, defining "heat wave" (using one word) as "a period of abnormally hot weather often defined with reference to a relative temperature threshold, lasting from two days to months. Heatwaves and warm spells have various and, in some cases, overlapping definitions". However they are defined, heat waves can be particularly deadly to poor people, such as those informal settlements in Odisha, India (Swain et al., 2019). The intensity of heat waves in South Asia is linked to human-caused climate change (Ullah et al., 2022), on the basis of which humanity has been warned of worse to come.

Another aspect of storm weather is lightning. Lightning strikes can kill and injure people through burns, stopping the heart, and nerve damage, especially to the brain. They interfere with electricity transmission and start infrastructure and vegetation fires. In the U.S., nearly one-fifth of lightning deaths are suffered by farmers and ranchers, as they work outdoors. Men are four times more likely to be struck by lightning than women (CDC, 2022). Thus, there are evident patterns of sex-based vulnerability. Yet the US experiences only about 20 lightning deaths per year, whereas Bangladesh can experience several hundred annually who are principally agricultural workers (Holle et al., 2019), again indicating the role of vulnerability.

Cyclonic storms with the regional names hurricane, typhoon, and cyclone cause damage with high wind, storm surge in coastal areas, and flooding. They can damage or destroy housing, crops, water supplies, sanitation and health care infrastructure, and educational facilities. They are a complex hazard with wind and different types of flooding often intersecting with multiple types of vulnerabilities (see the Box 5.4).

BOX 5.4 Cyclone Mocha in 2023

Cyclone Mocha crossed the Bay of Bengal beginning on 11 May 2023. The Global Disaster Alert and Coordination System (GDACS) issued a red alert on 17 May, citing "a high humanitarian impact based on the maximum sustained wind speed, exposed population and vulnerability" (GDACS, 2023). UNICEF alerted donors and first responders that "3.2 million people within the areas impacted by the cyclone are most vulnerable and likely to have humanitarian needs . . . [including] an estimated: 1.12 million children" (UNICEF, 2023). The storm made landfall just North of Sittwe, the capital of Myanmar's Rakine state. Winds were measured at 280 km/h, enough to make it a category five storm. Mocha brought dangerous winds, downpours, storm surge, and inland flooding. 40–50 deaths were reported among the 100,000 people affected in western Myanmar and 430,000 affected in Bangladesh (OCHA, 2023). The latter includes many Rohingya, living as refugees from Myanmar in several camps near Cox's Bazar, Bangladesh.

Recent events in this region provide insight into the kinds of root causes and dynamic pressures discussed in Chapter 1. These combined produce acutely unsafe conditions for the Rohingya, Bangladeshis, and Burmese (UNICEF, 2023). The cyclone came in the wake of armed conflict: the genocide by Myanmar's government against the Rohingya. Floods exposed unexploded shells and land mines, adding an additional hazard. Water supply, sanitation, and health infrastructure had already been damaged during the genocide, which was further battered by Cyclone Mocha. UNICEF (2023) reported that "the latrine superstructures in Rakhine camps have been substantially damaged and, in several locations, flooding is reported to have rendered the facilities dysfunctional. Similarly, heavy damage to water supply infrastructure has been reported, and the storm surge flooded water supply ponds, making the water saline" (UNICEF, 2023). All this increases the risk of outbreaks of water-related diseases, including cholera.

Living with Weather

An old maxim runs, "There is no such thing as bad weather, only bad clothing". But weather can be dangerous for people who do not have access to suitable clothing, equipment, shelter, food, water, indoor temperature control, and more. As always, cost and access to resources is a major constraint on reducing risk, in this case, weather-related. Harris et al. (1992), who designed public buildings, including schools for tornado-prone US areas, suggested that willingness to build better tornado resistance into designs is diminished by extra costs that reduce their chances of being awarded contracts. Instead, contractors have to aim for the lowest cost possible. To overcome this difficulty, measures such as having a tornado-safe room could be explicitly mentioned in tenders, and tornado-safety aspects could be included in building codes and planning regulations. These measures might not be feasible for some of the highly vulnerable dwellings, notably mobile homes, in which people of lower socio-economic status tend to live (Sutter & Simmons, 2010). They also tend to have less access to weather information and warnings, along with less opportunity to respond to warnings, such as no safe tornado shelter nearby.

Knowledge of expected or current weather, and the opportunity to respond to warnings, are essential for living with weather. Forecasts via news and social media are important sources. The accuracy of the information and the precision regarding location are major challenges. If too many people are told that a tropical cyclone will make landfall near them, then evacuation might be difficult. Mass evacuation nearly always involves deaths and injuries due to traffic collisions. As Hurricane Rita made a beeline for Texas in 2005, dozens perished in the heat and humidity when highway evacuation traffic stalled. If too few people are told about an impending storm, then casualties preventable through evacuation could result. Poor communication by forecasters can increase risk. As Typhoon Haiyan (Yolanda) bore down on the Philippines in 2013, forecasts were impressively accurate and precise, providing opportunities to evacuate. The warnings were not heeded for numerous reasons, one of which was that dangers from storm surge were not articulated in a way the population understood.

Living with weather means not only having the information and being able to interpret it but also being able to act on it. Evacuation in advance of a storm is not feasible without transportation, accommodation at one's destination, and the ability to pause employment without hardship. In 2005, many low-income residents of New Orleans, Louisiana, had no access to transportation and had to remain to face Hurricane Katrina. Highly localized tornado warnings can be issued 10–30 minutes beforehand in many cases, which is ample time to reach a tornado shelter – provided one exists and assuming that its location suits everyone's mobility and that people feel safe getting to and staying in the shelter. All these provisos involve vulnerabilities and abilities.

Climate

The climate is a statistical calculation of weather determined over decades by natural and human factors (IPCC, 2021–2022).

Natural Climate Change

The Earth's climate changes naturally and always has done. A stable or consistent local or planetary climate has never existed, and the climate system is mathematically chaotic (Annan & Hargreaves, 2004). Climate does not repeat a specific trajectory, although it typically stays within certain limits, for instance, of average global temperature. Challenges emerge in projecting future climates because small differences in the initial inputs can lead to a wide variety of outcomes. As such, directly from the definition of "climate", the idea of "climate chaos" is the norm, while "climate breakdown" cannot occur because climate is a statistical calculation that cannot break down per se.

Over epochs, sections of the Earth's surface called tectonic plates move and morph, in the process of continental drift (Chapter 6). Topography, bathymetry, and sea levels rise and fall as the sizes of continents and oceans wax and wane. These factors affect air currents, ocean currents, and heat exchange, meaning that weather changes and, in turn, climate changes. Another source of natural climate change is the Earth shifting its orbit around the Sun and so changing the energy received from the Sun.

For instance, early in the Eocene Epoch, about 53–49 million years ago, atmospheric carbon dioxide (CO_2) concentrations were four to six times the value today, and the mean global temperature was 10–18°C warmer than in the late 19th century (IPCC, 2021–2022). Ice Ages have brought ice sheets over 1 km thick to the mid-latitudes (Dyke et al., 2002). With that amount of water frozen, sea levels ended up 120 m lower than today. The rate of change of climate into and out of Ice Ages is typically assumed to occur over centuries, but major changes can occur within decades. Speculate how humanity might today deal with the onset of or emergence from an Ice Age within the span of a few generations!

Human-Caused Climate Change

Human activity influences the climate. Some connections are localized, such as cities augmenting air temperature and wind speed through the effects of urban heat islands and wind tunnels. Changing ecosystems through deforestation or reforestation influences the local climate. Microclimates can be created by releasing a steady flow of heated water, building greenhouses, or letting solid waste pile up. Besides these local influences, human actions have been causing a rise in global mean air temperature. Human activities release gasses, such as carbon dioxide and methane (CH_4), called "greenhouse gasses", which trap the sun's heat and warm the planet.

A principal activity emitting greenhouse gasses is the combustion of fossil fuels – mainly coal, oil, and natural gas – to power vehicles and generate electricity. Fossil fuels are also used to produce plastics. Other sources of greenhouse gas emissions are livestock farming and solid waste landfills, both of which release methane. Global warming is exacerbated by human activities that reduce the uptake from the atmosphere of greenhouse gasses. Land use management, including the destruction or protection of ecosystems, has a significant role in this regard. For instance, grasslands as pasturelands can be net greenhouse gas emitters or absorbers, depending on how they and the livestock are managed (Chang et al., 2021).

Greenhouse gas emissions that trap heat have had multiple impacts. Local air temperature and humidity are generally increasing globally. Bodies of water absorb the heat, raising their temperature and impacting land-ocean-air interactions, also bringing changes to the weather. Water above 4°C decreases in density as it gets warmer, so water volume expands for a given mass. As the oceans heat up, their volume increases and the sea level rises. Ice and snow melting adds further water mass and volume to the oceans. Sea-level rise increases the salt in coastal freshwater and inundates coastlines. Bodies of water also absorb carbon dioxide from the atmosphere which combines with water to form carbonic acid (H_2CO_3), increasing the water's acidity and modifying ecosystems.

Human-caused climate changes appear across many overlapping space and time scales, with numerous positive and negative feedbacks. The frequency, intensity, duration, location, and spatial extent of weather characteristics, such as rainfall rate and wind speed, change accordingly. People's lack of access to political and livelihood resources then turns such weather and climate change hazards into disasters. As established in Chapter 1, human actions, especially using and abusing power, create vulnerabilities and expose vulnerable people to hazards. Human actions also shape and modify hazards. Nevertheless, care should be taken with attribution of specific weather observations to the effects of human-caused climate (Herring et al., 2020), for instance:

1. Cyclonic storms typically increase in intensity but not frequency. The frequency of tropical cyclones (hurricanes, cyclones, and typhoons) is projected to decrease (Knutson et al., 2020), while frequency changes to polar lows and medicanes (cyclonic storms in the Mediterranean) vary seasonally and geographically (Landgren et al., 2019; Romero & Emanuel, 2017).
2. Flood (freshwater and saltwater, flash and slow-rise) frequency is much more affected by local human activity on the ground – such as engineering, land use, and urban development – than by weather.
3. Drought parameters are challenging to analyze, given how many droughts emerge from (over)use and (mis)management of water rather than from changes to the climate (Glantz & Katz, 1977; Wilhite & Glantz, 1985).
4. As with droughts, vegetation fires are significantly affected by human management of ecosystems and species (Aponte et al., 2016).
5. Impacts on regional climate variabilities have wide knowledge gaps and suggest inconsistent conclusions, with ENSO being said to weaken (Callahan et al., 2021) and strengthen (Cai et al., 2021) due to human-caused climate change.

Weather involving increased heat-humidity due to human-caused climate could have huge, adverse impacts on morbidity, mortality, livelihoods, food, and available freshwater (Romanello et al., 2021). As climate warms and becomes more humid over time, decreasing cold-related weather impacts on mortality and morbidity do not look likely to offset heat-humidity impacts, and this includes exacerbation of weather-related drought (Breshears et al., 2021). Bearing in mind the uncertainty of such forecasts, other potential catastrophic effects of human-caused climate change could include the following:

1. Sea-level rise. While sea-level rise from the immediate melting of snow, ice, and permafrost along with thermal expansion is problematic, much more impactful would be the melting of the Antarctic and/or Greenland ice sheets. Such melting would mean starting now to prepare for a major reconfiguration of world coastlines, which could happen over the next several centuries (Clark et al., 2016; Lowry et al., 2021; Noël et al., 2021).
2. Ocean acidification. The impacts are not fully known. Coral reefs could be especially hit hard when combined with higher sea surface temperatures and accelerating sea-level rise, although corals in their modern form have existed for over 200 million years (Stolarski et al., 2011). Coral reefs have shown remarkable survivability through diverse climates, but this does not guarantee their survival through human-caused climate change.
3. Ocean currents changing. Major shifts are possible in the ocean currents transporting heat and water mass around the world (e.g., Boers, 2021). The possible impacts are not fully mapped out. The slowing of the flow bringing warm water to Europe's higher latitudes, colloquially called the Gulf Stream, could offset some of that continent's warming as air temperatures would increase. Around the world, regional climates could change substantively and rapidly with this thermohaline circulation shift.
4. Permafrost and methane hydrate changes. Methane hydrate frozen in place in permafrost would be released in great quantities if permafrost were to thaw. There are scenarios in which substantial and rapid greenhouse gas release occurs due to such thawing, leading to accelerated human-caused climate change (Ruppel & Kessler, 2016).

Living with Climate Change – If Possible

In places where climate change threatens human survival, reducing vulnerability could mean leaving the areas affected, referred to as "managed realignment". In other words, where climate change merely alters the weather, but life and livelihoods are still possible, then responses should address vulnerabilities, not the environmental changes per se. With appropriate governance, resource allocation, political power distribution, education, and actions across many scales, the changing climate should have minimal impact on disaster risk.

Nevertheless, both natural and human-caused climate change could challenge the viability of human life and livelihoods. Irrespective of possibilities for a new Ice Age or a return to prehistoric high temperatures, the current projections for anthropogenic climate change require both adaptation (vulnerability reduction) and climate change mitigation which means decreasing greenhouse gas emissions and increasing their uptake (IPCC, 2021–2022). For the latter, significant debates swirl around suggestions to reduce the amount of solar radiation reaching the Earth and attempts to affect the amount of solar radiation leaving the Earth. They are often referred to as "geoengineering", highlighting grandiose constructions, such as placing giant filters, mirrors, or lenses, between the Earth and the Sun to control the amount of radiation arriving.

Currently, the efforts to affect the amount of solar radiation leaving the Earth by altering greenhouse gas composition in the atmosphere are more feasible. The current necessity is to decrease the concentration of such gasses by reducing emissions and increasing their uptake. The latter is termed "carbon capture and storage" or "carbon capture and sequestration" (CCS). Reducing greenhouse gas emissions requires weaning humanity off fossil fuels, by reducing energy demand and shifting toward supplies that do not deplete irreplaceable resources. Examples of alternative energy sources include localized solar, wind, tidal, and hydropower along with local mechanisms, such as heat pumps and osmosis energy in saltwater. This shift requires resources, particularly for the long-haul shift away from established energy infrastructure. The money is available from the subsidies given to fossil fuels, which annually are hundreds of billions of dollars in direct subsidies (IEA, 2021) and trillions of dollars in indirect subsidies (Coady et al., 2019). Thus, the resources exist, but they need to be reallocated to stopping, rather than creating, climate change.

Greenhouse gas absorption can be increased by managing ecosystems and materials. Ecosystem management means planting and maintaining species that absorb the gasses, although care is needed to select species and ecosystems typical for local areas to avoid causing significant environmental harm by introducing non-native species. Materials management refers to using materials that draw down and store greenhouse gasses. Timber is an example, so trees could be planted and harvested for structures, including multi-story buildings, followed by planting more in the logged locations. Bamboo is already used in this way in some places (Zhao & Qiu, 2023).

Irrespective of how quickly climate change mitigation is enacted, human-caused climate change impacts are already evident (e.g., on weather, Herring et al., 2020), and many more are locked in for coming decades (IPCC, 2021–2022). More conscious and rapid adaptation and mitigation are required, partly because humanity is experiencing weather it has never before seen, such as the heat-humidity combinations discussed earlier (IPCC, 2021–2022).

Two overarching lessons emerge from addressing climate change through the lenses of both mitigation and adaptation (Kane & Shogren, 2000). First, do not address climate change in isolation. Without considering wider and deeper contexts, other problems might result. A cautionary example is US and EU governmental support for diesel vehicles in the late 1990s and early 2000s on the premise that they produce fewer greenhouse gas emissions than petrol vehicles, only to have this stance reversed two decades later through bans on diesel vehicles because their particulate emissions are health hazards, and they also emit significant amounts of greenhouse gasses (Tong & Azevedo, 2020; Wilkinson & Haines, 2015; California Government, 2001). The second overarching lesson is that climate change, including that caused by humans, does not need to be invoked in order to support the required actions. All actions related to climate change have non-climate change justification for their pursuit. Climate change adaptation is easily supported as disaster risk reduction. Climate change mitigation is easily justified because fossil fuels are expensive, finite, and polluting. The transition to renewable energy sources and the development of energy-conserving technologies and materials are potential sources of livelihoods and employment while saving money overall by withdrawing subsidies from the fossil fuel industry.

Interconnection Among Water, Weather, Climate, and Beyond

As alluded to throughout this chapter, water, weather, and climate are not isolated from each other or other environmental phenomena. Rain, or lack thereof, affects slope stability and so mass movements. Sea-level rise and ice sheets melting redistribute mass around the Earth, which could trigger earthquakes and volcanic eruptions (McGuire, 2010). Other chapters address these hazards, noting that water, weather, and climate are just some of the influences that must be considered. Understanding risk requires understanding them all and their interconnections.

References Cited

Ahmed, B., Kelman, I., Kamruzzaman, M., Mohiuddin, H., Rahman, M. M., Das, A., Fordham, M., & Shamsudduha, M. (2019). Indigenous people's responses to drought in northwest Bangladesh. *Environmental Development, 29*, 55–66. https://doi.org/10.1016/j.envdev.2018.11.004

Albert, J. S., Destouni, G., Duke-Sylvester, S. M., Magurran, A. E., Oberdorff, T., Reis, R. E., Winemiller, K. O., & Ripple, W. J. (2021). Scientists' warning to humanity on the freshwater biodiversity crisis. *Ambio, 50*, 85–94. https://doi.org/10.1007/s13280-020-01318-8

Allan, W. (1977). *The African husbandman*. Greenwood.

Annan, J. D., & Hargreaves, J. C. (2004). Efficient parameter estimation for a highly chaotic system. *Tellus A: Dynamic Meteorology and Oceanography, 56*(5), 520–526. https://doi.org/10.3402/tellusa.v56i5.14438

Aponte, C., de Groot, W. J., & Wotton, B. M. (2016). Forest fires and climate change: Causes, consequences and management options. *International Journal of Wildland Fire, 25*(8), i–ii. http://dx.doi.org/10.1071/WFv25n8_FO

Arunotai, N. (2008). Saved by an old legend and a keen observation: The case of Moken sea nomads in Thailand. In R. Shaw, N. Uy, & J. Baumwoll (Eds.), *Indigenous knowledge for disaster risk reduction: Good practices and lessons learnt from the Asia-Pacific region* (pp. 73–78). UNISDR Asia.

Baxter, S. (2008). *Flood*. Gollancz.

Bertenthal, A. (2021). Scaling the baseline: Technicalities and environmental regulation in Owens Valley, California. *Law and Policy, 43*(1), 30–50. https://doi.org/10.1111/lapo.12162

Boers, N. (2021). Observation-based early-warning signals for a collapse of the Atlantic meridional overturning circulation. *Nature Climate Change, 11*, 680–688. https://doi.org/10.1038/s41558-021-01097-4

Bolonkin, A. A. (2011). Floating cities. In S. D. Brunn (Ed.), *Engineering earth: The impacts of megaengineering projects* (pp. 967–983). Springer.

Breshears, D. D., Fontaine, J. B., Ruthrof, K. X., Field, J. P., Feng, X., Burger, J. R., Law, D. J., Kala, J., & St. Hardy, J. G. E. (2021). Underappreciated plant vulnerabilities to heat waves. *New Phytologist, 231*(1), 32–39. https://doi.org/10.1111/nph.17348

Burlinson, A., Davillas, A., & Law, C. (2022). Pay (for it) as you go: Prepaid energy meters and the heat-or-eat dilemma. *Social Science and Medicine, 315*, Article 115498. https://doi.org/10.1016/j.socscimed.2022.115498

Cai, W., Santoso, A., Collins, M., Dewitte, B., Karamperidou, C., Kug, J.-S., Lengaigne, M., McPhaden, M. J., Stuecker, M. F., Taschetto, A. S., Timmermann, A., Wu, L., Yeh, S.-W., Wang, G., Ng, B., Jia, F., Yang, Y., Ying, J., Zheng, X.-T., . . . Zhong, W. (2021). Changing El Niño–Southern oscillation in a warming climate. *Nature Reviews Earth & Environment, 2*, 628–644. https://doi.org/10.1038/s43017-021-00199-z

California Government. (2001). *Health effects of diesel exhaust*. California Office of Environment Health and Hazards Assessment. https://oehha.ca.gov/air/health-effects-diesel-exhaust

Callahan, C. W., Chen, C., Rugenstein, M., Bloch-Johnson, J., Yang, S., & Moyer, E. J. (2021). Robust decrease in El Niño/Southern oscillation amplitude under long-term warming. *Nature Climate Change, 11*, 752–757. https://doi.org/10.1038/s41558-021-01099-2

CBD (Convention on Biodiversity). (2003). *The status and trends in biodiversity of Inland water ecosystems*. Convention on Biodiversity.

CDC (U.S. Center for Disease Control and Prevention). (2022). *Natural disasters and severe weather. Lightning strike victim data*. https://www.cdc.gov/disasters/lightning/victimdata.html

Chang, J., Ciais, P., Gasser, T., Smith, P., Herrero, M., Havlík, P., Obersteiner, M., Guenet, B., Goll, D. S., Li, W., Naipal, V., Peng, S., Qiu, C., Tian, H., Viovy, N., Yue, C., & Zhu, D. (2021). Climate warming from managed grasslands cancels the cooling effect of carbon sinks in sparsely grazed and natural grasslands. *Nature Communications, 12*, 118. https://doi.org/10.1038/s41467-020-20406-7

Clark, P. U., Shakun, J. D., Marcott, S. A., Mix, A. C., Eby, M., Kulp, S., Levermann, A., Milne, G. A., Pfister, P. L., Santer, B. D., Schrag, D. P., Solomon, S., Stocker, T. F., Strauss, B. H., Weaver, A. J., Winkelmann, R., Archer, D., Bard, E., Goldner, A., . . . Plattner, G.-K. (2016). Consequences of twenty-first-century policy for multi-millennial climate and sea-level change. *Nature Climate Change, 6*, 360–369. https://doi.org/10.1038/nclimate2923

Coady, D., Parry, I., Le, N.-P., & Shang, B. (2019). *Global Fossil fuel subsidies remain large: An update based on country-level estimates*. IMF working paper. Fiscal Affairs Department, International Monetary Fund.

Comité d'Information Sahel. (1975). *Qui se nourrit de la famine en Afrique*? F. Maspero.

Copans, J. (Ed.). (1975). *Sécheresses et famines du Sahel*. F. Maspero.

Criss, R. E., & Shock, E. L. (2001). Flood enhancement through flood control. *Geology, 29*(10), 875–878. https://doi.org/10.1130/0091-7613(2001)029%3C0875:FETFC%3E2.0.CO;2

Dancause, K. N., Chan, C. W., Arunotai, N. H., & Lum, J. K. (2009). Origins of the Moken sea gypsies inferred from mitochondrial hypervariable region and whole genome sequences. *Journal of Human Genetics, 54*, 86–93. https://doi.org/10.1038/jhg.2008.12

Davis, M. (2006). *City of quartz: Excavating the future in Los Angeles* (New ed.). Verso.

Doberstein, B., Fitzgibbons, J., & Mitchell, C. (2019). Protect, accommodate, retreat or avoid (PARA): Canadian community options for flood disaster risk reduction and flood resilience. *Natural Hazards, 98*, 31–50. https://doi.org/10.1007/s11069-018-3529-z

Dudgeon, D. (2019). Multiple threats imperil freshwater biodiversity in the Anthropocene. *Current Biology, 29*(19), R960–R967. https://doi.org/10.1016/j.cub.2019.08.002

Dyke, A. S., Andrews, J. T., Clark, P. U., England, J. H., Miller, G. H., Shaw, J., & Veillette, J. J. (2002). The Laurentide and Innuitian ice sheets during the last glacial maximum. *Quaternary Science Reviews, 21*, 9–31. https://doi.org/10.1016/S0277-3791(01)00095-6

Etkin, D. (1999). Risk transference and related trends: Driving forces towards more mega-disasters. *Environmental Hazards, 1*, 69–75. https://doi.org/10.1016/S1464-2867(00)00002-4

Etkin, D., & Maarouf, A. (1995, February 11–14). An overview of atmospheric natural hazards in Canada. In D. Etkin (Ed.), *Proceedings of a tri-lateral workshop on natural hazards, Sam Jakes Inn, Merrickville, Canada* (pp. 1–92). Environment Canada.

FAA (U.S. Federal Aviation Administration). (2015). *Windshear detection services.* https://www.faa.gov/air_traffic/weather/wsds

Fleuret, A. (1986). Indigenous responses to drought in sub-Saharan Africa. *Disasters, 10*(3), 224–229. https://doi.org/10.1111/j.1467-7717.1986.tb00592.x

GDACS (Global Disaster Alert and Coordination System). (2023). *Overall red alert tropical cyclone for MOCHA-23 in Myanmar.* https://www.gdacs.org/meteo.aspx?eventid=1000970&episodeid=1&eventtype=TC

Glantz, M. H. (1976). *The politics of natural disaster: The case of the Sahel drought.* Praeger.

Glantz, M. H., & Katz, R. W. (1977). When is a drought a drought? *Nature, 267,* 192–193. https://doi.org/10.1038/267192a0

Goudie, A. (2006). *The human impact on the natural environment: Part, present and future.* Wiley-Blackwell.

Harris, H. W., Mehta, K. C., & McDonald, J. R. (1992). Taming Tornado Alley. *Civil Engineering, 62,* 77–78.

Herring, S. C., Christidis, N., Hoell, A., Hoerling, M. P., & Stott, P. A. (Eds.). (2020). Explaining extreme events of 2018 from a climate perspective. *Bulletin of the American Meteorological Society, 101,* S1–S141. https://doi.org/10.1175/BAMS-ExplainingExtremeEvents2018.1

Herzberg, I. (Executive Producer), Licht, A. (Executive Producer), Mueller, J. (Executive Producer), & Reynolds, K. (Director). (1995). *Waterworld* [Motion Picture]. Universal Pictures.

Holle, R. L., Dewan, A., Said, R., Brooks, W. A., Hossain, M. D., & Rafiuddin, M. (2019). Fatalities related to lightning occurrence and agriculture in Bangladesh. *International Journal of Disaster Risk Reduction, 41,* 101264. https://doi.org/10.1016/j.ijdrr.2019.101264

IEA. (2021). *Fossil fuel subsidies database: Fossil fuel consumption subsidies for select countries, 2010–2020.* IEA (International Energy Agency).

IPCC. (2021–2022). *Sixth assessment report.* IPCC (Intergovernmental Panel on Climate Change).

Jackson, J. (2001). Living with earthquakes: Know your faults. *Journal of Earthquake Engineering, 5*(special issue 1), 5–123. https://doi.org/10.1080/13632460109350530

Kaji-o'grady, S., & Raisbeck, P. (2005). Prototype cities in the sea. *The Journal of Architecture, 10*(4), 443–461. https://doi.org/10.1080/13602360500285641

Kane, S., & Shogren, J. F. (2000). Linking adaptation and mitigation in climate change policy. *Climatic Change, 45,* 75–102. https://doi.org/10.1023/A:1005688900676

Knutson, T., Camargo, S. J., Chan, J. C. L., Emanuel, K., Ho, C.-H., Kossin, J., Mohapatra, M., Satoh, M., Sugi, M., Walsh, K., & Wu, L. (2020). Tropical cyclones and climate change assessment: Part II: Projected response to anthropogenic warming. *Bulletin of the American Meteorological Society, 101*(3), E303–E322. https://doi.org/10.1175/BAMS-D-18-0194.1

Konikow, L., & Kendy, E. (2005). Groundwater depletion: A global problem. *Hydrogeology Journal, 13*(1), 317–320. https://doi.org/10.1007/s10040-004-0411-8

Landgren, O. A., Batrak, Y., Haugen, J. E., Støylen, E., & Iversen, T. (2019). Polar low variability and future projections for the Nordic and Barents Seas. *Quarterly Journal of the Royal Meteorological Society, 145,* 3116–3128. https://doi.org/10.1002/qj.3608

Lehner, B., & Dölla, P. (2004). Development and validation of a global database of lakes, reservoirs and wetlands. *Journal of Hydrology, 296,* 1–22. https://doi.org/10.1016/j.jhydrol.2004.03.028

Lewin, P. A., Fisher, M., & Weber, B. (2012). Do rainfall conditions push or pull rural migrants: Evidence from Malawi. *Agricultural Economics, 43*(2), 191–204. https://doi.org/10.1111/j.1574-0862.2011.00576.x

Lowry, D. P., Krapp, M., Golledge, N. R., & Alevropoulos-Borrill, A. (2021). The influence of emissions scenarios on future Antarctic ice loss is unlikely to emerge this century. *Communications Earth & Environment, 2,* 221. https://doi.org/10.1038/s43247-021-00289-2

Lucas, G. (Executive Producer and Director). (1999). *Star Wars: Episode I - The Phantom Menace* [Motion Picture]. Lucasfilm.

McGuire, B. (2010). Potential for a hazardous geospheric response to projected future climate changes. *Philosophical Transactions of the Royal Society A: Mathematical, Physical and Engineering Sciences, 368,* 2317–2345. https://doi.org/10.1098/rsta.2010.0080

Mortimore, M. (2010). Adapting to drought in the Sahel: Lessons for climate change. *Wiley Interdisciplinary Reviews: Climate Change, 1*(1), 134–143. https://doi.org/10.1002/wcc.25

Noël, B., van Kampenhout, L., Lenaerts, J. T. M., van de Berg, W. J., & van den Broeke, M. R. (2021). A 21st century warming threshold for sustained greenland ice sheet mass loss. *Geophysics Research Letters, 48*(5), Article e2020GL090471. https://doi.org/10.1029/2020GL090471

OCHA. (2023, May). *Tropical cyclone Mocha.* OCHA (United Nations Office for the Coordination of Humanitarian Affairs). https://reliefweb.int/disaster/tc-2023-000069-mmr

Riffat, S., Powell, R., & Aydin, D. (2016). Future cities and environmental sustainability. *Future Cities and Environment, 2,* 1. https://doi.org/10.1186/s40984-016-0014-2

Romanello, M., McGushin, A., Di Napoli, C., Drummond, P., Hughes, N., Jamart, L., Kennard, H., Lampard, P., Solano Rodriguez, B., Arnell, N., Ayeb-Karlsson, S., Belesova, K., Cai, W., Campbell-Lendrum, D., Capstick, S., Chambers, J., Chu, L., Ciampi, L., Dalin, C., . . . Hamilton, I. (2021). The 2021 report of the lancet countdown on health and climate change. *The Lancet, 398*(10311), 1619–1662. https://doi.org/10.1016/S0140-6736(21)01787-6

Romero, R., & Emanuel, K. (2017). Climate change and hurricane-like extratropical cyclones: Projections for North Atlantic polar lows and medicanes based on CMIP5 models. *Journal of Climate, 30*(1), 279–299. https://doi.org/10.1175/JCLI-D-16-0255.1

Ruppel, C. D., & Kessler, J. D. (2016). The interaction of climate change and methane hydrates. *Reviews of Geophysics, 55*(1), 126–168. https://doi.org/10.1002/2016RG000534

Smucker, T. A., & Wisner, B. (2008). Changing household responses to drought in Tharaka, Kenya: Vulnerability, persistence and challenge. *Disasters, 32*(2), 190–215. https://doi.org/10.1111/j.1467–7717.2007.01035.x

Stolarski, J., Kitahara, M. V., Miller, D. J., Cairns, S. D., Mazur, M., & Meibom, A. (2011). The ancient evolutionary origins of Scleractinia revealed by azooxanthellate corals. *BMC Evolutionary Biology, 11,* 316. https://doi.org/10.1186/1471-2148-11-316

Sutter, D., & Simmons, K. (2010). Tornado fatalities and mobile homes in the United States. *Natural Hazards, 53,* 125–137. https://doi.org/10.1007/s11069-009-9416-x

Swain, S., Bhattacharya, S., Dutta, A., Pati, S., & Nanda, L. (2019). Vulnerability and adaptation to extreme heat in Odisha, India: A community based comparative study. *International Journal of Environmental Research and Public Health, 16*(24), 5065. https://doi.org/10.3390/ijerph16245065

Szöllösi-Nagy, A., & Zevenbergen, C. (Eds.). (2005). *Urban flood management.* A.A Balkema.

Tobin, G. A. (1995). The Levee love affair: A stormy relationship. *Water Resources Bulletin, 31*(3), 359–367. https://doi.org/10.1111/j.1752-1688.1995.tb04025.x

Tong, F., & Azevedo, I. M. L. (2020). What are the best combinations of fuel-vehicle technologies to mitigate climate change and air pollution effects across the United States? *Environmental Research Letters, 15*(7), 074046. https://doi.org/10.1088/1748-9326/ab8a85

Ullah, S., You, Q., Chen, D., Sachindra, D. A., AghaKouchak, A., Kang, S., Li, M., Zhai, P., & Ullah, W. (2022). Future population exposure to daytime and nighttime heat waves in South Asia. *Earth's Future, 10*(5), Article e2021EF002511. https://doi.org/10.1029/2021EF002511

UNICEF. (2023). *UNICEF Myanmar humanitarian situation report no. 5 (Cyclone MOCHA).* https://reliefweb.int/report/myanmar/unicef-myanmar-humanitarian-situation-report-no-5-cyclone-mocha-19-may-2023

Wilhite, D. A., & Glantz, M. H. (1985). Understanding the drought phenomenon: The role of definitions. *Water International, 10*(3), 111–120. https://doi.org/10.1080/02508068508686328

Wilkinson, P., & Haines, A. (2015). Diesel in the dock. *BMJ, 351,* Article h5415. https://doi.org/10.1136/bmj.h5415

Wisner, B. (1995). Bridging "expert" and "local" knowledge for counter-disaster planning in urban South Africa. *GeoJournal, 37,* 335–348. https://doi.org/10.1007/BF00814014

Zhao, J. C., & Qiu, H.-X. (2023). Seismic performance assessment of a multi-story bamboo frame structure. *Advances in Bamboo Science, 2,* 100011. https://doi.org/10.1016/j.bamboo.2022.100011

Chapter Six

Earthly Hazards

Earthquakes and Volcanoes

Geological Background to Our Planet's Dynamism

Throughout history, without understanding that the Earth's crust is broken into plates that move, people have taken advantage of hot springs and artesian wells, enjoying abundant harvests grown on rich volcanic soils. In tectonically active regions, they also feared sudden violent movements of what was usually taken for granted as *terra firma*. Our ancestors thanked supernatural beings for the harvests taken from the slopes of dormant volcanoes. Likewise, the ancients explained earthquakes and volcanoes as the result of the activities of gods or subterranean creatures.

A Chinese emperor nearly 2,000 years ago commissioned the invention of a device to detect earthquakes hundreds of kilometers distant in order to know if the gods were calm or agitated. The precursor of the seismograph was the product of both science and superstition. The composition of the earth and earthquakes have provoked the curiosity of thinkers for centuries. A follower of Aristotle, Theophrastes wrote a volume on stones based partly on observations he made visiting mines near Athens. The Roman philosopher Pliny the Elder and medieval Persian scholar Avicenna shared these interests. Fossils produced much speculation about the earth's history by theologians in the 17th and 18th centuries.

Secular thinkers, such as Immanuel Kant, also wondered about fossils' origins and the earth's composition. However, none of these studies revealed the nature of the crust, mantle, and deeper layers until the early 1900s, when Alfred Wegener put forth the idea of continental drift (Wegener, 1912). Wagener based his theory on various factors, such

DOI: 10.4324/9781003292814-6

as the positioning of continents, similar fossil plants and animals, and rock layers across oceans, as well as spotting cold climate signs near the equator and warm climate signs near the poles. The theory suggested that all the continents were once a single entity known as Pangaea, but over time, they drifted apart (Caamaño-Alegre, 2018).

The theory of continental drift was replaced by plate tectonics due to its limited explanation of the mechanism and understanding of drift patterns. Plate tectonics, developed between 1962 and 1968, suggests that massive rock slabs, known as tectonic plates, support the continents, constantly moving and interacting with each other (Figure 6.1). Even today, the continents continue to shift, with the most dynamic sites of tectonic activity being seafloor spreading zones and giant rift valleys. During seafloor spreading, the molten rock rises from within the Earth and adds new oceanic crust to the edges of the old.

It is fundamental to understand the composition of our planet to gain insight into its origins and how it operates today. Scientists have made progress in identifying the building blocks of Earth and the processes involved in its development. However, the exact makeup of the Earth's core remains a mystery, with only iron and nickel alloys confirmed. Some experts suspect other elements could be present, such as oxygen, sulfur, or silicon, based on density differences observed at the inner and outer core boundary (McDonough & Sun, 1995).

The Earth has a layered structure, with the core made of molten metallic alloy and covered by a thick layer of silicates and oxides. The "Moho" or Mohorovic discontinuity, located near the core-mantle boundary (CMB), played a significant role in the dynamic systems of the core and mantle throughout their 4.5 billion years of evolution. The CMB, denser than the Earth's surface and has much higher temperatures than found near the surface, was a key factor in this process (Lay et al., 1998).

The surface of the Earth is composed of plates separated by faults. These faults come in three types: normal, thrust/reverse, and strike-slip (Figure 6.2). If a fault is classified as active, it has moved within the past 2 million years. While Bird's (2003) study found 52 plates, it has since been suggested by Harrison (2016) that there are 159 plates. The forces at the edges of these plates cause mantle flow, with slab pull being a common force. Understanding both the lithosphere and asthenosphere is key to comprehending tectonic activity. The lithosphere, which is the outermost rocky layer, is made strong by lower temperatures. The asthenosphere, on the other hand, is a softer, more fluid layer that drives the movement of the plates.

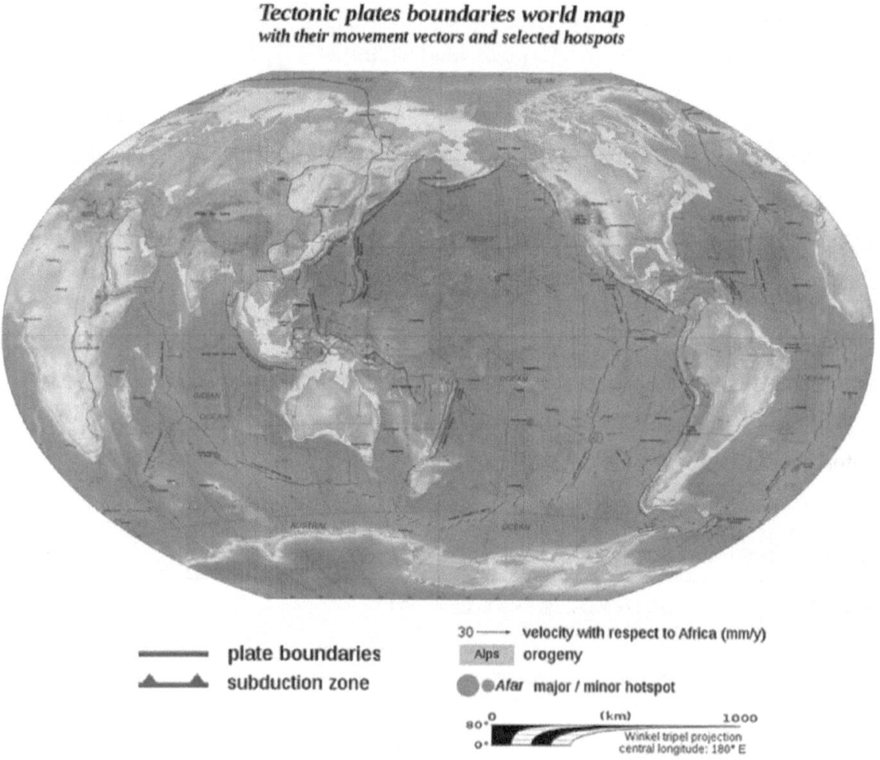

FIGURE 6.1 Map of Earth's principal tectonic plates

Source: "Eric Gaba for Wikimedia Commons", CC BY-SA 3.0, https://commons.wikimedia.org/w/index.php?curid=66499832

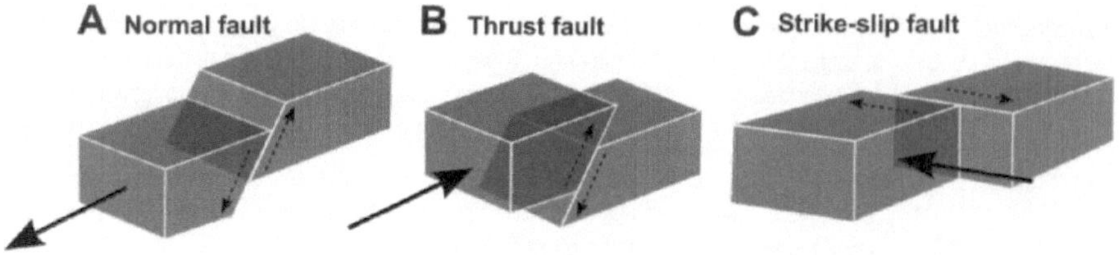

FIGURE 6.2 Types of faults

Source: Jesús Gómez Fernández, CC BY-SA 3.0 <http://creativecommons.org/licenses/by-sa/3.0/>, via Wikimedia Commons

Earthquakes

Earthquakes can happen when stress builds up and exceeds the strength of the local materials. They cause seismic waves and can damage already weakened structures with aftershocks. Some earthquakes happen slowly (Kanamori, 2004), while others involve sudden and unstable slip events (Dieterich, 1978). These events can be faster than the tectonic stress release and are caused by fault asperities.

Different types of earthquakes occur at varying depths in the earth's crust. Shallow earthquakes occur from 0 to 70 km, while intermediate-depth earthquakes occur from 70 to 300 km (Green & Houston, 1995; Scholz, 2019). Deep earthquakes are similar to shallow ones in rupture behavior but have different physical processes. Their characteristics depend on factors like slab temperature, depth, and magnitude (Zhan, 2020).

Earthquakes come in different forms, such as rapid fault line slipping and volumetric or planar expansion near volcanoes. Scientists classify them based on their slip rate, which ranges from km/s to mm/yr (McCaffrey, 2021). For more information on earthquakes and related phenomena, refer to Table 6.1.

TABLE 6.1 Types of earthquakes and related phenomena

Earthquakes and related phenomena: types and definition	
Tectonic earthquakes	The stick-slip behavior of faults causes tectonic earthquakes due to stress buildup. When the maximum shear stress is exceeded, the fault snaps back, releasing high-frequency seismic waves. They are the most common type of earthquakes.
Tectonic tremors	Tectonic tremors are a new type of seismic disturbance first observed in Japan in 2002. Like volcanic tremors, they last for days to weeks with low-amplitude, non-harmonic background disturbances. These tremors are often associated with slow-slip events and are believed to be part of the same process as tectonic earthquakes.
Volcanic earthquakes	Volcanic earthquakes are like tectonic ones, caused by rock failure along faults. However, they are usually smaller because of the limited fault surface around volcanoes.
Volcanic tremors	Earth disturbances known as volcanic tremors are possibly caused by fluids such as water or magma moving within the earth. These signals tend to be harmonic and can last for an extended period, although they typically have low amplitude and produce minimal crustal deformation compared to tectonic earthquakes.
Slow earthquakes	Slow-slip earthquakes can occur in less rigid materials, like accretionary prisms in subduction zones. They can still generate seismic waves and sometimes create unexpectedly large tsunamis for the earthquake's magnitude.
Low-frequency earthquakes (LFEs)	In volcanic settings and subduction zones, there have been observations of low-frequency earthquakes that generate much lower frequencies than regular tectonic earthquakes. Fluid interactions may cause these low frequencies.
Slow-slip events (SSEs)	Slow-slip events spread a few millimeters daily, much slower than earthquakes but faster than tectonic creep. Seismometers cannot detect them, but geodetic instruments have found them on subduction zones, such as Kilauea and the San Andreas fault.
Post-seismic slip events	After a big earthquake, there can be post-seismic slip due to continued fault slip at slow speeds. This can last years and cause a slip equivalent to the main earthquake, but most occur without generating seismic waves. Typically, after-slip events usually decrease over time.

(*Continued*)

TABLE 6.1 (Continued)

Earthquakes and related phenomena: types and definition	
Mantle relaxation	After major earthquakes, the stress in the Earth's crust and mantle can shift, causing the mantle to flow viscously. This flow can lead to a long-period deformation of the crust above, which is sometimes mistaken for post-seismic slip. This phenomenon is known as mantle relaxation.
Induced earthquakes	Human activities, such as mining, drilling, and water impoundment behind dams, can modify the stress at shallow depths in the Earth, resulting in induced earthquakes. These earthquakes are particularly hazardous, as they occur close to infrastructures and are shallow in depth.
Fault morphology	Faulting is how the Earth's crust deforms. Crustal faults are enduring evidence of past earthquakes, and through geological research, it is possible to learn more about these seismic events. Long-term deformation rates in continental areas can be estimated by analyzing slip rates and fault lengths.

Source: McCaffrey (2021)

Measuring Earthquakes and Their Impact

Seismologists use many different scales to measure earthquakes (Table 6.2). There is the intensity scale, which measures the strength of the shaking, and the magnitude scale, which measures the energy released. The Mercalli and Modified Mercalli scales, which were developed in the early 1900s (Mercalli, 1902; Wood & Neumann, 1931), illustrate how earthquakes impact people, structures, and the environment, helping to understand their effects.

During an earthquake, the energy released is measured by magnitude. Two types of surface waves, namely, Love and Rayleigh, can cause retrograde motion or sideways shaking. Body waves, including P-waves that move quickly through rock and S-waves that move in a transverse motion, are slower than P-waves (Gates & Ritchie, 2006). To measure the magnitude of earthquakes, the Richter scale, also known as the Gutenberg-Richter scale, was developed by examining the amplitude of waves recorded by seismographs (Richter, 1935). This base-10 logarithmic scale means that each order of magnitude is ten times more intense than the previous one. For instance, a level 2 earthquake is ten times more intense than a level 1, while a level 3 is 100 times greater.

Although the Richter scale is effective, it has limitations, as it only applies to specific frequency and distance ranges. Creating new scales, such as body wave magnitude (*Mb*) and surface wave magnitude (*Ms*), using data from seismograph stations worldwide has been an attempt to address these limitations. However, it is important to note that all three scales (*ML*, *Mb*, and *Ms*) have some limitations, as each type of seismic wave differs in energy, leading to variations in earthquake magnitude depending on the type of wave used (Yanovskaya, 2008; Shearer, 2019; Pulinets & Ouzounov, 2018) (Table 6.3).

TABLE 6.2 Modified Mercalli scale

Intensity	Shaking	Description/Damage
I.	***Not felt***	Not felt, except by a very few, under especially favorable conditions.
II.	***Weak***	Felt only by a few people at rest, especially on the upper floors of buildings.
III.	***Weak***	Felt quite noticeable by people indoors, especially on the upper floors of buildings. Many people do not recognize it as an earthquake. Standing motor cars may rock slightly. Vibrations are like the passing of a truck. Duration estimated.
IV.	***Light***	Felt indoors by many and outdoors by few during the day. At night, some are awakened. Dishes, windows, and doors are disturbed; walls make cracking sounds. Sensations are like a heavy truck striking a building. Standing motor cars are rocked noticeably.
V.	***Moderate***	Felt by nearly everyone; many awakened. Some dishes and windows were broken. Unstable objects overturned. Pendulum clocks may stop.
VI.	***Strong***	Felt by all, many frightened. Some heavy furniture moved; a few instances of fallen plaster. Damage slight.
VII.	***Very strong***	Damage is negligible in buildings of good design and construction, slight to moderate in well-built ordinary structures; considerable damage in poorly built or badly designed structures; some chimneys broken.
VIII.	***Severe***	Damage is slight in specially designed structures, considerable damage in ordinary substantial buildings with partial collapse. Damage is great in poorly built structures. Fall of chimneys, factory stacks, columns, monuments, and walls. Heavy furniture overturned.
IX.	***Violent***	Damage is considerable in specially designed structures; well-designed frame structures are thrown out of plumb. Damage is great in substantial buildings, with partial collapse. Buildings shifted off foundations.
X.	***Extreme***	Some well-built wooden structures are destroyed; most masonry and frame structures are destroyed with foundations. Rails bent.

Source: https://www.usgs.gov/media/images/modified-mercalli-intensity-scale

TABLE 6.3 Differences in the scales of magnitude

Region	Date	M_b	M_S	M_W	M_o
Chile	22 May 1960	–	8.3	9.5	2000
Alaska	28 March 1964	–	8.4	9.2	820
Sumatra-Andaman	26 December 2004	6.2.	8.5	9.1.	680
Aleutian Islands	9 March 1957	–	8.2	9.1.	585
Aleutian Islands	4 February 1965	–	–	8.7	140
Sumatra	28 March 2005	7.2	8.4	8.6	105
Indonesia	19 August 1977	7.0	7.9	8.3	36
Hokkaido, Japan	25 September 2003	6.9	8.1	8.3	31
Shikotan, Kurile Islands	4 October 1994	7.4	8.1	8.2	30

(*Continued*)

TABLE 6.3 (Continued)

Region	Date	M_b	M_S	M_W	M_o
Bolivia	9 June 1994	6.9	–	8.2	26
Macquarie Ridge	23 December 2004	6.5	7.7	8.1	16

Source: Shearer (2019)
M_b body waves; M_s surface waves; M_w instantaneous magnitude; M_0 moment magnitude

TABLE 6.4 The Environmental Seismic Intensity (ESI) scale

Intensity	Shaking	Description/Damage
I.	***Not felt***	Only a very few people can feel it and only under especially favorable conditions.
II.	***Weak***	It was only felt by a few resting people, particularly on the higher floors of buildings.
III.	***Weak***	The earthquake was quite noticeable to people indoors, especially on the upper floors of buildings. Many people do not recognize it as an earthquake. Parked cars may rock slightly, and the vibrations were similar to those caused by the passing of a truck. The duration of the event is estimated.
IV.	***Light***	During the day, many people feel it indoors, and some feel it outdoors. At night, some are awakened by it. Dishes, windows, and doors are disturbed; walls make cracking sounds. The sensation is similar to a heavy truck striking a building, and standing motor cars are noticeably rocked.
V.	***Moderate***	Felt by almost everyone; many awakened. Some dishes and windows are broken, and unstable objects are overturned. Pendulum clocks may stop.
VI.	***Strong***	Felt by everyone, many are frightened. Some heavy furniture was moved, and there were a few instances of fallen plaster. The damage was slight.
VII.	***Very strong***	Damage is minimal in well-designed and constructed buildings, moderate in well-built ordinary structures, considerable in poorly constructed or designed buildings, and some chimneys are broken.
VIII.	***Severe***	Damage is minimal in specifically designed structures; substantial damage is seen in ordinary buildings with partial collapse. Poorly built structures suffer significant damage, such as falling chimneys, factory stacks, columns, monuments, and walls, with heavy furniture being overturned.
IX.	***Violent***	Damage is substantial in specifically designed structures; well-designed frame structures are displaced. Significant damage occurs in large buildings, with partial collapse. Buildings are shifted off foundations.
X.	***Extreme***	Some well-constructed wooden buildings are destroyed, while most masonry and frame structures are destroyed, including their foundations. The rails are bent.

Source: https://www.usgs.gov/media/images/modified-mercalli-intensity-scale

The International Union developed the Environmental Seismic Intensity (ESI) scale for *Quaternary Research* to assess the impact of earthquakes on the environment. ESI considers significant factors, such as surface faulting and tectonic uplift/subsidence, and secondary factors, such as landslides, ground cracks, liquefactions, displaced boulders, tsunamis, and hydrological anomalies (Table 6.4).

The relationship between intensity and magnitude scales is crucial in understanding the impact of earthquakes worldwide. Vulnerable communities residing in high seismicity areas are at risk, and it is imperative to implement various strategies to mitigate this exposure. Governments are responsible for preventing adverse consequences, and one of the most common mechanisms is the earthquake early warning system.

Earthquake Early Warning Systems (EEWSs)

Effective early warning systems (EWSs) require a functional social process to prevent harm from hazards. This process includes clear communication, informed decision-making, accurate risk assessment, customized approaches, and avoiding standardization that could hinder the integration of local expertise, which is crucial for each unique social context. Despite efforts to implement EWSs, the main challenge remains in overcoming the technical view of outsiders and providing expert information to the community (Garcia & Fearnley, 2012; Kelman & Glantz, 2014).

For warnings, striving to comprehend earthquake dynamics has proven to be a significant endeavor. Researchers have explored several potential precursors, including radon levels in local water, electromagnetic signals from Earth's crust, and unusual animal behavior. None of these findings has yet withstood statistical tests. Despite many promising leads, there has been limited progress in earthquake prediction (Geller, 1997; Hall, 2023).

Animal behavior is often regarded as a controversial earthquake signal. Many people believe that abnormal animal behavior is a precursor to an earthquake. However, Woith et al. (2018) conducted a study on the subject and found that it is unclear if this behavior is directly related to the nucleation process of an earthquake. They also noted that analyzing animal behavior is more challenging than measuring physical field parameters with instruments and sensors.

Seismologists use different approaches to analyze the Earth's crust for earthquake precursors and assess hazards (Pulinets & Ouzounov, 2018), including probabilistic and deterministic estimates. Probabilistic estimates express forecasts as a probability and require studying the area's historical seismicity and characterizing the geological tectonic factors on local or regional scales (Keilis-Borok, 1996). Deterministic estimates rely on physical laws relating to precursors and express predictions through alarms (Martinelli, 1998).

TABLE 6.5 Ground-based earthquake precursors

Precursor		References
Seismicity and extremely low-frequency acoustic emissions	Extremely low-frequency acoustic emissions before the main rupture at higher frequencies.	Mignan, 2008; Hong et al., 2018; De Santis et al., 2019; Ihmlé & Jordan, 1994
Lithospheric mechanical deformations	Mechanical deformations in the lithosphere.	Niu et al., 2008; Langbein et al., 2005
Variation of the groundwater level and composition	Variation in the groundwater level and composition.	Hayakawa et al., 1997; Koizumi et al., 1999
Gas exhalations	The release of gas, particularly radon or radioactive ions.	Khilyuk et al., 2000; Pulinets et al., 2003
Fluctuations of temperature	Temperature fluctuations are possibly linked to groundwater circulation, uplift changes, or vapor condensation on the surface.	Tramutoli et al., 2005
Propagation of acoustic gravity waves (AGW)	Acoustic gravity waves (AGW). A physical mechanism of seismo-ionospheric coupling, including AGW and radon exhalation, has been recently suggested.	Molchanov & Hayakawa, 2008; Rapoport et al., 2020
Fluctuation of electric and magnetic field components	Fluctuating electric and magnetic field components.	Uyeda et al., 2009; Han et al., 2014; Sorokin et al., 2020
Ground-based observations of ionospheric parameters	Monitoring ionospheric variables from the surface, including but not limited to ground-based observations of ionospheric parameters, such as total electron content (TEC), very low-frequency reflection height, whistler dispersion, etc.	Liu et al., 2004; Liu, 2009; Hayakawa et al., 1993, 1996, 1997; Rodger et al., 1999; Hobara & Parrot, 2005

Source: Conti et al. (2021)

Earthquake Precursors

Earthquake precursors indicate stress or strain that could lead to an earthquake. They should appear on multiple instruments or at different sites, and their size should increase as the earthquake approaches. While it is impossible to predict exactly when an earthquake will occur, these signs can indicate that one might be coming at a specific location (Pulinets & Ouzounov, 2018; Conti et al., 2021) (Table 6.5).

Earthquake and EEWSs Awareness and Education

It is crucial to create assessments of earthquake disaster risks while understanding seismicity and predicting earthquakes. Earthquake early warning systems (EEWSs) can estimate the size of an earthquake based on initial P-waves, which can provide warnings, perhaps in the range of 30–60 seconds, of potentially damaging S-waves. As seismic faulting mainly involves shear faulting, the initial P-wave poses little threat, while the later-arriving S-wave carries enough energy to cause damage. Using this characteristic of energy radiation during seismic faulting, it is possible to develop reliable early warning techniques suitable for today's societies with complex structures (Kanamori, 2004).

Some countries have earthquake early warning systems (EEWSs) that use seismometers to detect quakes and notify people (see Box 6.1), although cities nearer to the epicenter receive less warning time (Suárez, 2022). Addressing risk factors and promoting disaster awareness through education is essential, rather than relying on technology to prevent disasters.

BOX 6.1 People's Awareness and Knowledge, Experiencing EEWSs: Japan's EEWS on 11 March 2011 and SASMEX: The Mexican EEWS

Japan's earthquake early warning system on 11 March 2011

In Japan, two networks – Hi-net and JMA's Tsunami and Earthquake Observation and Monitoring system – can send real-time data, which is vital for the EEWS. The EEWS comprises ten systems that serve different purposes, such as the fire department, medical, and home electronics systems. Four EEWs transmit alerts using disaster prevention wireless, IP telephone, public mobile communications, and FM broadcasting systems. The effectiveness of the EEWs depends on how widely warnings are distributed and how many people benefit from them. The educational campaign for the EEWs is essential in informing the public about what actions to take when an alert is issued. A survey conducted by the JMA after the Tohoku earthquake found that 87% of respondents were familiar with the EEWS due to prior alerts. The survey also showed that people were aware of the system's limitations, but the educational campaign effectively explained the system's structure, and 78% understood the possibility of false alarms (Fujinawa & Noda, 2013).

Mexico's SASMEX earthquake early warning system on 19 September 2017

Mexico has a system called SASMEX that warns people of earthquakes. It has been working for 30 years and has sent 111 alerts to 25 million people. When the alerts are sent, people get a map that shows how strong the earthquake will be. This gives people up to 60 seconds to protect themselves before the earthquake arrives (Suárez, 2022). When the 2017 earthquake happened, it was hard to warn people close to it. To address this problem, education programs that explain the basics of earthquake dynamics should be implemented. Different sounds or countdowns could also warn people how close the earthquake is. Near to the epicenter, people might not have time to react. Developing an integrated perspective helps communities understand an earthquake is not the only determining factor in a disaster involving seismicity.

Sources: Fujinawa & Noda (2013); Suárez, (2022)

Apart from detecting earthquakes and educating people about EEWSs, focusing on reducing vulnerability is paramount. This involves creating policies that ensure safe construction practices by implementing appropriate building standards and codes. These policies should be permanent and robust to provide a secure and sustainable environment.

Building Codes

Ensuring compliance with construction and planning regulations in areas prone to seismic activity requires significant effort during the design and construction phases. Doing so can be hindered by corruption and ignorance, resulting in a lack of adherence to codes, rules, and regulations (Lewis, 2003). This issue is widespread in several countries, such as Türkiye (see Box 6.2) and Mexico, raising questions about the accountability of authorities responsible for civil protection and disaster risk management at various levels of government (Alcántara-Ayala et al., 2022). It is vital to uphold building and planning regulations, as any lack of integrity in professional conduct can result in severe consequences. Nevertheless, developers, engineers, architects, planners, and builders may fail to realize the significance of complying with standards, essential in protecting the interests of building owners, users, designers, and builders (Alcocer, 2023).

It is important to learn from past disasters involving earthquakes. Understanding seismic hazards' dynamics and analyzing societal dynamics are necessary to prevent future

BOX 6.2 The 7.8 Mw Earthquake of 6 February 2023: Türkiye and Syria

On 6 February 2023, at 4:17 A.M. local time, a 7.8 Mw earthquake struck Türkiye and Syria, followed by a second 7.5 Mw seism on the same day at 13:24 P.M. in Ekinozu/Kahramanmaras province. These earthquakes resulted in the devastating loss of over 55,000 lives and injured more than 108,000 people. At least 13.5 million individuals felt the impact of the disaster. The data from Türkiye's Ministry of Environment, Urbanization, and Climate Change (MMAUCC) reveals that a staggering 38,018 structures collapsed, 11,868 of which were made of concrete. Over 20,000 buildings require immediate demolition. Among the total number of structures in the region (2.14 million), nearly 15% sustained moderate or severe damage. The Office of Strategy and Budget estimates that the disaster will cost 103.6 billion US dollars. Of the total cost, 55% is attributed to the damage to housing. These earthquakes are the largest that the country has experienced since the 7.8 Mw tremor in 1939 (Alcocer, 2023).

Source: Alcocer (2023)

disasters. Thus, assessing earthquake risk based on diverse criteria requires careful consideration of various factors, including land use, resources, socio-economic dynamism and structure, earthquake systems dynamics, fault analysis, and wave amplification – with basic vulnerability parameters being most important: Who lives in the area, and what opportunities do they have to deal with earthquakes with a disaster happening? Reducing vulnerability to earthquakes presents significant challenges. Accountability mechanisms must be improved to address this, particularly in housing and infrastructure planning development – such as building and maintaining without corruption. This will prevent the creation of new risks and help manage existing risk.

Volcanoes

Throughout history, there have been various interpretations of what constitutes a volcano. Borgia et al. (2010) defined it as a geological setting that involves three interconnected components: magma, eruption, and edifice. If one of these elements can be verified, and the existence or potential of the other two can be inferred, then it can be classified as a volcano. Németh (2010) suggested that most volcanoes are in four settings: rift volcanoes (Walker, 1999), subduction volcanoes, hot-spot volcanoes (Jellinek

& Manga, 2004), and intraplate volcanoes (Johnson, 1989). Alexander von Humboldt first defined volcanism in 1845 as the essence of all the reactions of a planet against its crust and surface, and this notion has since incorporated other elements, such as temperature.

Types of Volcanoes

One classification of volcanoes provides four types: cinder cones, composite, shield, and lava domes. Cinder cones are steep hills made of volcanic debris, while composite volcanoes are conical and made of lava and pyroclastic deposits. Shield volcanoes are shaped like shields and made of basaltic lava flows. Lava domes are small formations of thick, slow-moving lava associated with explosive eruptions (De Silva & Lindsay, 2015).

Calderas and submarine volcanoes are volcanic structures formed by explosive or effusive eruptions. When a large explosive or effusive eruption causes subsidence, calderas create a wide basin with elevated rims, while submarine volcanoes are formed underwater. Caldera eruptions can vary in composition and volume, lasting from days to months or hours to days (Branney & Acocella, 2015).

Submarine volcanoes have a significant role in creating oceanic crust, seamounts, and island arcs. They produce deposits through various processes and interact with the surrounding water, which releases gas and particles. This interaction affects magma ascent and cooling. When there are explosive submarine eruptions, gas and particles are released into the environment, impacting magma ascent, fragmentation, particle cooling, and dispersal. On the other hand, submarine-effusive eruptions create primary volcaniclastic deposits through nonexplosive processes (White et al., 2015a, 2015b).

Effusive and Explosive Eruptions

Volcanic eruptions can be classified as explosive or effusive. Explosive eruptions occur when gas trapped in high-viscosity magma builds up and is released, causing ash plumes and pyroclastic flows. Effusive eruptions involve low-viscosity magma, which allows gas to escape easily and produce different types of lava flows. Understanding volcanic eruptions is crucial for society to increase knowledge of volcanic hazards and disaster risks. Recent advancements have led to investigating different lava types generated by basalt volcanoes and forming volcanic lava domes, which are important in understanding the subtle transformations between effusive and explosive activity (Self, 2006).

Studying volcanoes at mid-ocean ridges reveals how intrusive rocks create new oceanic crust. Seamounts formed in different tectonic settings are also important. Research on basaltic volcanic fields shows varying magmatic output and composition. Understanding flood basalts and large igneous provinces is crucial due to massive supervolcano eruptions that can cover up to 100,000 square kilometers over lengthy time periods (Sigurdsson et al., 2015).

For studying volcanic eruptions, the use of high-resolution data and modeling has allowed a better understanding of the various types of volcanic hazards and their potential impact on society. Valuable insights have been gained by analyzing the nature and products of explosive volcanism, the mechanisms behind volcanic explosions, and the dynamics of transport and deposition. These insights can help us prepare better for future volcanic events (Sigurdsson et al., 2015).

Hazards of Volcanism, Monitoring, and Warning Systems

Volcanoes are classified based on their activity level, which is significant in hazard dynamics. Active volcanoes have erupted at least once in the last 10,000 years and can either be actively erupting or dormant. Extinct volcanoes are categorized as young, recent, old, or eroded. Both dormant and young volcanoes have the potential to become active (Szakács, 1994) and pose volcanic hazards when their processes and landforms have the potential to affect societies. Direct and chained secondary volcanic hazards occur during or immediately after an eruption episode (Table 6.6).

Analyzing various factors in real time is crucial when monitoring volcanic activity. These factors include eruption onset, duration, plume/jet height, mass eruption rate, erupted volume/mass, exit velocity, energy, and ratios between different types of energy. After an eruption, parameters like the volume/mass of different volcanic products, plume height, mass eruption rate, duration, and grain size distribution need to be considered (Bonadonna et al., 2016).

Studying active and potentially active volcanoes requires real-time analysis, observation, monitoring, and post-eruption assessments. These tasks are challenging due to high costs and increased challenges over time. National volcano monitoring and warning systems should be tailored to meet the specific needs of each context, sometimes local, sometimes global, sometimes in-between, and sometimes multiple or all scales.

The VEI scale measures the explosive potential or actuality of an eruption. It considers various factors to determine the magnitude of an eruption, which directly correlates with the volume of magma released. The logarithmic relationship between the volume of

TABLE 6.6 Direct and concatenated volcanic hazards

Volcanic Hazards	Direct volcanic hazards	*Fall processes:* Tephra falls Ballistic projectiles ***Flowage processes:*** Pyroclastic flows, surges Laterally directed blasts Debris avalanches Primary debris flows (lahars) Floods (jökulhlaups) Lava flows ***Other processes:*** Phreatic explosions Edifice collapses Volcanic gasses and acid rains Near-vent and plume lightning
	Chained secondary volcanic hazards	Earthquakes and ground movements Volcanogenic tsunamis Flooding and alterations to the drainage and pattern of river flows Secondary debris flows (lahars) Secondary pyroclastic flows Long-distance volcanic ash dispersion
Volcanic Hazards	Direct volcanic hazards	*Fall processes:* Tephra falls Ballistic projectiles ***Flowage processes:*** Pyroclastic flows, surges Laterally directed blasts Debris avalanches Primary debris flows (lahars) Floods (jökulhlaups) Lava flows ***Other processes:*** Phreatic explosions Edifice collapses Volcanic gasses and acid rains Near-vent and plume lightning
	Chained secondary volcanic hazards	Earthquakes and ground movements Volcanogenic tsunamis Flooding and alterations to the drainage and pattern of river flows Secondary debris flows (lahars) Secondary pyroclastic flows Long-distance volcanic ash dispersion

(*Continued*)

TABLE 6.6 (Continued)

Volcanic Hazards	Direct volcanic hazards	*Fall processes:* Tephra falls Ballistic projectiles
		Flowage processes: Pyroclastic flows, surges Laterally directed blasts Debris avalanches Primary debris flows (lahars) Floods (jökulhlaups) Lava flows
		Other processes: Phreatic explosions Edifice collapses Volcanic gasses and acid rains Near-vent and plume lightning
	Chained secondary volcanic hazards	Earthquakes and ground movements
		Volcanogenic tsunamis
		Flooding and alterations to the drainage and pattern of river flows
		Secondary debris flows (lahars)
		Secondary pyroclastic flows
		Long-distance volcanic ash dispersion

Sources: Scott (1989); Tilling (2005)

ejected material and the VEI ensures that even minor eruptions can be distinguished from colossal ones (Newhall & Self, 1982).

Volcanoes as Factual Hazard

The Krakatau volcanic eruption that happened in 1883 lasted several months and its intensity was measured at a 6 on the volcanic explosivity index (VEI). The eruption caused massive tsunamis and explosions (Self, 1992), and the volume of magma released was around 10 km^3. The surrounding regions and far-off areas were severely affected by the disaster. The eruption's aftermath was marked by extensive ash deposits and other volcanic material that caused widespread devastation to the surrounding areas. The resulting tsunamis led to a staggering loss of life, estimated at more than 36,000 people (Carey et al., 2000).

A super-eruption, a rare volcanic phenomenon that deposits significant amounts of pyroclastic materials like ash and rock fragments, can have a massive impact on the global climate and the environment due to the release of volcanic and sulfur gasses (Sparks et al., 2005; Lowenstern et al., 2006; Self, 2006). Although these remarkable eruptions only occur once every 100,000–200,000 years, slightly smaller VEI 7 eruptions happen more frequently (Self, 2006).

Yellowstone in North America has the potential to produce some rare, powerful, explosive eruptions (Self, 2006). Christiansen (2001) identified a massive eruption known as the VEI 8.7 Huckleberry Ridge Tuff. According to USGS (n.d.), the ash from super-eruptions covered a huge chunk of North America around 2 million and 600,000 years ago. Oppenheimer (2002) also mentioned that the ash from Toba's eruption in Sumatra, Indonesia, around 74,000 years ago was found in places as far away as northern India, Pakistan, and the south Indian Ocean.

It is important to compare different volcanoes and their risks since they can greatly impact society (Box 6.3). Scandone et al. (2016) devised a way to measure this, called the volcanic risk coefficient (VRC). It considers things like the biggest possible eruption, how long it has been since the last one, and how many people are at risk. The highest VRC is 17, for huge volcanoes like Toba or Yellowstone, which could affect all life on Earth. More research is needed to make sure everything is considered, especially social (including economic) factors. Some recent studies look at Southeast Asia's exposure to volcanic hazards (Jenkins et al., 2022) and combine social vulnerability and hazard simulations (Liu et al., 2020).

BOX 6.3 The Impact of 15 January 2022 at the Hunga Tonga-Hunga Ha'apai Submarine Volcano, Tonga

In January 2022, a colossal eruption occurred at the Hunga Tonga-Hunga Ha'apai submarine volcano. Studies by Zhao et al. (2022) estimated the volcanic explosivity index (VEI) to be between 5 and 6. Proud et al. (2022) reported that the volcanic plume reached an unprecedented height of 57 km, causing a fast and significant tsunami, as noted by Gusman et al. (2022), at great distances from the source. The volcanic cloud traveled to the southern hemisphere's polar region in a few weeks. It reached the northern hemisphere polar region by April 2022, resulting in the highest optical depth of stratospheric aerosols since the Pinatubo eruption in 1991, as reported by Taha et al. (2022). Adushkin et al. (2022) also suggested that the eruption caused global variations in the electric and magnetic fields. The volcanic eruption and accompanying tsunami caused significant damage and loss in Tonga, with four fatalities, over 84,000 people affected, and damage estimated at US$117 million. Tonga lost both its international and domestic submarine fiber cables. The agricultural sector was severely affected, with volcanic activity accounting for 80% and the tsunami 20%, as was the tourism sector. The hazards also impacted transportation, power, and freshwater.

The Adverse Impacts of Volcanic Hazards

Exposure to volcanic risk over time would depend on the nature and depth of the overall understanding and connectedness of people and volcanoes. The level of human and scientific capabilities, the complexity of an economy, and skewed development leading to shaping people's vulnerability would determine the disaster risk level.

Considering this approach to the systemic risk associated with the diverse manifestations of volcanic activity, a list of the adverse impacts of volcanic hazards has been identified. It should be noted that the list is indicative but not by any means not limiting. Likewise, it is acknowledged that to understand the social construction of systemic risk fully, it is necessary to analyze, understand, and address all spheres associated with vulnerability and exposure in addition to the hazards (see Table 6.7).

TABLE 6.7 The adverse impact of volcanic hazards

Direct impact	*Social impact:* Human losses, injured, missing, displaced, and affected people (assets, livelihoods, and health). *Physical damage to capital assets:* The collapse of roofs, transport, transportation networks, communications, digital connectivity linked to fiber-optic cables, power outages, telecommunications blackouts, irrigation, factories, machinery, industrial plants, other types of infrastructure, houses, hospitals, schools, and public buildings, crops, livestock, and the fishing industry. *Aviation:* Volcanic ash affects the airframe and electrical, hydraulic, and navigation systems, compromising high-bypass jet-turbine engines that run at temperatures over the melting point of many minerals and silicates. *Environmental:* Contamination of water, crops, and air by toxic volcanic ash. Contamination of water supplies, block, and damage to water and sewage systems. Nutritional stress is induced by damage or destruction of crops, trees, and other vegetation.
Concatenated legacy	*Social impact:* Psychological effects; respiratory health issues. Loss of jobs. Medical expenses and lost productivity derived from increased disease, injury, and incidence of death.

(*Continued*)

TABLE 6.7 (Continued)

	Environmental: Volcanic sulfur emissions. Adverse influence on the radiation budget, surface temperatures, and circulation patterns of the atmosphere due to sulfuric acid aerosol clouds. Depletion of stratospheric ozone in temperate-polar latitudes, including a relationship to increased levels of damaging ultraviolet light. Volcanic aerosols affect stratospheric stability and influence tropospheric dynamics, particularly temperature change on land masses (eruptions in tropical latitudes produce winter warming in northern hemisphere continents but cooling in summer). Production of greenhouse warming by large emissions of CO2. Destruction or modification of existing vegetation through burial, burning, or blowdowns of vegetation. The toxicity can also result in the slow death of plants and trees. Soil changes and increases in carbon dioxide. Impossibility to use agricultural lands for long periods of time.
	Long-term financial impact: Damage to the flow of goods and services, including lower output from damaged or destroyed assets and infrastructure. Increased prices due to supply shortages reflecting the damage to goods, transport, and marketing infrastructure. Loss of earnings due to damage to marketing infrastructure (roads, ports, airports) and to lower effective demand and the costs associated with using more expensive inputs after the destruction of cheaper usual sources of supply. Costs of both medical expenses and lost productivity derived from increased incidence of disease, injury, and death. Impact on net inflation. Government monetary and fiscal policy shifts to contain the effects of increased disaster-induced inflation or to finance additional government expenditure. The influence of the local, regional, and nationwide distribution of investment resources associated with potential implications for widening regional inequalities. Demands on limited budgetary resources. Additional unplanned expenditure or the partial redeployment of previously allocated funds for relief and rehabilitation operations. Increasing budgetary pressures for local and federal governments. Closure of airports, schools, hospitals, and other vital infrastructures. Short-, medium-, and long-term impacts on overall economic performance, such as deterioration in trade and government budget balances and increased indebtedness, as well as the impact on income distribution or the scale and incidence of poverty. Long-term debt when provided in the form of soft loans instead of grants.

Sources: Vogelmann et al. (1992); Self (2005)

Benefits of Volcanic Activity

Major benefits can emerge from living in a volcanic area, notably linked to a sense of belonging and closeness to nature. Cultural practices and traditional beliefs regarding volcanic activity should be recognized and integrated into programs to facilitate education and communication (Box 6.4). Moreover, it is essential to emphasize the interdependence and mutual support among different types of volcanic activity and environmental legacies.

Volcanoes play a significant role in various aspects of daily life, including cultural, social (including economic), and environmental vitality (see Box 6.5). One example is a wide range of industrial uses of volcanic materials. Geothermal energy reservoirs and their uses support development in nearby places. Proper use of volcanic resources, particularly tourism, fosters livelihoods and, if used properly, local social progress. Volcano-derived soils are fertile and provide for profitable farming, requiring sound knowledge of volcanic activity's benefits. A gain in this knowledge will reduce disaster risks and preserve a more harmonious relationship between humanity and the environment. In addition, volcanoes have given meaning to the puzzle of combining feelings, ideas, awareness, truths, histories, stories, imaginations, dreams, and thoughts into the art of philosophy and poetry.

Quite recently, a 2,500-year-old ancient city with 6,000 house mounds, roads, and canals has been discovered in the Amazon region in eastern Ecuador. The abundant soil

BOX 6.4 Indigenous Strategies for Disaster Risk Reduction in the Baliau Community, Papua New Guinea

The Baliau community is situated on the northwest coast of Manam Island, around 16 km from the Bogia District of Madang Province in Papua New Guinea. It is affected by the nearby Manam volcano and so uses various disaster risk reduction strategies, including building methods, social linkages, land use planning, and food and environmental strategies. These strategies are often embedded in their culture and daily life, helping them live with the active volcano. During a large volcanic eruption in 2004, an evacuation was necessary but poorly executed, causing difficulties for the people. Indigenous communities should be involved in decision-making, as they possess valuable knowledge. Using both Indigenous and scientific knowledge can help communities like Baliau. The sustainable livelihoods approach can analyze the effects of living near a volcano, leading to a more informed approach. Consulting with communities and authorities can establish a culturally sensitive process for evacuation and rehabilitation that considers all factors together.

Source: Mercer & Kelman (2010)

generated by a nearby volcano facilitated the city's high population density. Although it is believed that the city was eventually deserted due to a volcanic eruption, this discovery offers fresh perspectives on ancient civilizations' ingenuity and ability to adapt to their environment (Rostain et al., 2024) (Box 6.6).

BOX 6.5 Living with Mount Bromo, East Java, Indonesia

Mount Bromo, standing at 2,392 m above sea level, is a highly active volcano in East Java, Indonesia. Its location is within the vast Tengger caldera, which has a rich history of volcanic activity. Since 1804, Mount Bromo has erupted on at least 56 occasions. The most recent eruption occurred in 2010 and lasted for nine months. Despite the area's high level of volcanic hazards, the Tengger people have chosen to settle in this region. They have learned to cope with the dangers and found ways to benefit from the various physical, spiritual, social, and cultural opportunities that arise within the human-volcanic system. Overall, they have shown remarkable adaptability and resourcefulness in the face of volcanic activity. Through time, the Tengger people have developed five significant cultural adaptations from living within the Bromo human volcano system. These include heightened resilience and capacity to recover from disasters, a strong attachment to the place and hazard knowledge, a source of social and moral order within the community, a means to frame and voice dissent, and a catalyst for change processes.

Source: Bachri et al. (2015)

BOX 6.6 Discovering the Connections Between a Lost City and a Volcano in the Amazonian Ecuador

Recent archaeological research has uncovered a pre-Hispanic urban system in the Upano Valley of Amazonian Ecuador, featuring clusters of monumental platforms, plazas, streets, and a complex road system connecting different urban centers. The area was occupied from 500 BCE to between 300 and 600 CE. The Kilamope and Upano people were sedentary agrarian societies that heavily occupied the valley, where fertile volcanic soils still allow up to three harvests per year. The Sangay volcano's activity had a profound impact on pre-Hispanic communities.

Previous studies showed continuous cone growth and two major flank failures during the Pleistocene, resulting in debris avalanches that spread as far as 60 km from the volcano. Thick, hummocky deposits left behind by these avalanches were the foundation for human settlements during the late Holocene period. Furthermore, analyses indicate that the area experienced significant explosive eruptions in the past. Despite the hypothesis that the Upano culture ended abruptly after a massive eruption around 400 to 600 CE, the inconsistency of recent radiocarbon dates calls this theory into question.

Source: Rostain et al. (2024)

Conclusions

Earthquakes and volcanoes have been part of the evolving human landscapes since the beginning of civilization. They are an integral part of livelihoods, in which synergies and multiple benefits are constantly shaped. Living with earthquakes and volcanoes requires maximizing synergies and minimizing trade-offs. Achieving this requires a deep understanding of the different types of eruptions and seismic activity and associated hazards in order to implement effective instrumentation and monitoring systems around the world. While addressing the underlying causes of disaster risk requires reducing vulnerability (including exposure), effective communication and exchange also play a crucial role in preventing misunderstandings and the spreading of false information.

References Cited

Adushkin, V. V., Rybnov, Y. S., & Spivak, A. A. (2022). Geophysical effects of the eruption of Hunga–Tonga–Hunga–Ha'apai volcano on January 15, 2022. *Doklady Earth Science, 504*, 362–367. https://doi.org/10.1134/S1028334X22060034

Alcántara-Ayala, I., Rodríguez-Velázquez, D., & Lucatello, S. (2022). Under the rubble: Disaster risk management and accountability after the Mexico City earthquake of September 19, 2017. *Natural Hazards, 120*, 9111–9134. https://doi.org/10.1007/s11069-022-05752-7

Alcocer, S. M. (2023, May 9). Apuntes sobre los sismos en Turquía y Siria (Notes on the earthquakes in Turkey and Syria). *Letras Libres*. https://letraslibres.com/ciencia-tecnologia/sergio-alcocer-apuntes-sismos-turquia-siria/

Bachri, S., Stötter, J., Monreal, M., & Sartohadi, J. (2015). The calamity of eruptions, or an eruption of benefits? Mt. Bromo human–volcano system a case study of an open-risk perception. *Natural Hazards and Earth System Sciences, 15*(2), 277–290. https://doi.org/10.5194/nhess-15-277-2015

Bird, P. (2003). An updated digital model of plate boundaries. *Geochemistry, Geophysics, Geosystems, 4*(3). https://doi.org/10.1029/2001GC000252

Bonadonna, C., Cioni, R., Costa, A., Druitt, T., Phillips, J., Pioli, L., Andronico, D., Harris, A., Scollo, S., Bachmann, O., Bagheri, G., Biass, S., Brogi, F., Cashman, K., Dominguez, L., Dürig, T., Galland, O., Giordano, G., Gudmundsson, M., Hort, M., Höskuldsson, A., Houghton, B., Komorowski, J. C., Küppers, U., Lacanna, G., Le Pennec, J. L., Macedonio, G., Manga, M., Manzella, I., de' Michieli Vitturi, M., Neri, A., Pistolesi, M., Polacci, M., Ripepe, M., Ross, E., Scheu, B., Sulpizio, R., Tripoli, B., Valade, S., Valentine, G., Vidal, C., & Wallenstein, N. (2016). MeMoVolc report on classification and dynamics of volcanic explosive eruptions. *Bulletin of Volcanology, 78*(84), 1–12. https://doi.org/10.1007/s00445-016-1071-y

Borgia, A., Aubert, M., Merle, O., & Van Wyk De Vries, B. (2010). What is a volcano? In E. Cañón-Tapia & A. Szakács (Eds.), *What is a volcano?* (pp. 1–9). Geological Society of America Special Paper 470. https://doi.org/10.1130/2010.2470(01)

Branney, M., & Acocella, V. (2015). Calderas. In *The encyclopedia of volcanoes* (2nd ed., pp. 299–315). https://doi.org/10.1016/B978-0-12-385938-9.00016-X

Caamaño-Alegre, M. (2018). Drift theory and plate tectonics: A case of embedding in geology. *Foundations of Science, 23*, 17–35. https://doi.org/10.1007/s10699-016-9505-8

Carey, S., Sigurdsson, H., Mandeville, C., & Bronto, S. (2000). Volcanic hazards from pyroclastic flow discharge into the sea: Examples from the 1883 eruption of Krakatau, Indonesia. In F. W. McCoy & G. Heiken (Eds.), *Volcanic hazards and disaster in human antiquity* (pp. 1–14). Geological Society of America. https://doi.org/10.1130/0-8137-2345-0.1

Christiansen, R. L. (2001). *The quaternary and Pliocene Yellowstone Plateau volcanic fields of Wyoming, Idaho, and Montana* (Vol. 729). US Department of the Interior, US Geological Survey. https://doi.org/10.3133/pp729G

Conti, L., Picozza, P., & Sotgiu, A. (2021). A critical review of ground based observations of earthquake precursors. *Frontiers in Earth Science, 9*, 676766. https://doi.org/10.3389/feart.2021.676766

De Santis, A., Abbattista, C., Alfonsi, L., Amoruso, L., Campuzano, S. A., Carbone, M., Cesaroni, C., Cianchini, G., De Franceschi, G., De Santis, A., Di Giovambattista, R., Marchetti, D., Martino, L., Perrone, L., Piscini, A., Rainone, M. L., Soldani, M., Spogli, L., & Santoro, F. (2019). Geosystemics view of earthquakes. *Entropy, 21*(4), 412. https://doi.org/10.3390/e21040412

De Silva, S., & Lindsay, J. M. (2015). Primary volcanic landforms. In *The encyclopedia of volcanoes* (2nd ed., pp. 273–297). https://doi.org/10.1016/B978-0-12-385938-9.00015-8

Dieterich, J. H. (1978). Time-dependent friction and the mechanics of stick-slip. *PAGEOPH, 116*, 790–806. https://doi.org/10.1007/BF00876539

Fujinawa, Y., & Noda, Y. (2013). Japan's earthquake early warning system on 11 March 2011: Performance, shortcomings, and changes. *Earthquake Spectra, 29*(1), 341–368. https://doi.org/10.1193/1.4000127

Garcia, C., & Fearnley, C. J. (2012). Evaluating critical links in early warning systems for natural hazards. *Environmental Hazards, 11*(2), 123–137. https://doi.org/10.1080/17477891.2011.609877

Gates, A. E., & Ritchie, D. (2006). *Encyclopedia of earthquakes and volcanoes*. Infobase Publishing.

Geller, R. J. (1997). Earthquake prediction: A critical review. *Geophysical Journal International, 131*(3), 425–450. https://doi.org/10.1111/j.1365-246X.1997.tb06588.x

Green, H. W., & Houston, H. (1995). The mechanics of deep earthquakes. *Annual Review of Earth and Planetary Sciences, 23*(1), 169–213. https://doi.org/10.1146/annurev.ea.23.050195.001125

Gusman, A. R., Roger, J., Noble, C., Wang, X., Power, W., & Burbidge, D. (2022). The 2022 Hunga Tonga-Hunga Ha'apai volcano air-wave generated tsunami. *Pure and Applied Geophysics, 179*(10), 3511–3525. https://doi.org/10.1007/s00024-022-03154-1

Hall, S. (2023). What Turkey's earthquake tells us about the science of seismic forecasting. *Nature, 615*(7952), 388–389. https://doi.org/10.1038/d41586-023-00685-y

Han, P., Hattori, K., Hirokawa, M., Zhuang, J., Chen, H., Febriani, F., Yamaguchi, H., Yoshino, C., Liu, Y., & Yoshida, S. (2014). Statistical analysis of ULF seismomagnetic phenomena at Kakioka, Japan, during 2001–2010. *Journal of Geophysical Research: Space Physics, 119*(6), 4998–5011. https://doi.org/10.1002/2014JA019789

Harrison, C. G. (2016). The present-day number of tectonic plates. *Earth, Planets and Space, 68*(1), 1–14. https://doi.org/10.1186/s40623-016-0400-x

Hayakawa, M., Kawate, R., Molchanov, O. A., & Yumoto, K. (1996). Results of ultra-low-frequency magnetic field measurements during the Guam earthquake of 8 August 1993. *Geophysical Research Letters, 23*(3), 241–244. https://doi.org/10.1029/95GL02863

Hayakawa, M., Molchanov, O. A., Ondoh, T., & Kawai, E. (1997). *On the precursory signature of Kobe earthquake on VLF subionospheric signals*. EEE International Symposium on Electromagnetic Compatibility, Beijing, 72–75. https://doi.org/10.1109/elmagc.1997.617080

Hayakawa, M., Yoshino, T., & Morgounov, V. (1993). On the possible influence of seismic activity on the propagation of magnetospheric whistlers at low latitudes. *Physics of the Earth and Planetary Interiors, 77*(1–2), 97–108. https://doi.org/10.1016/0031-9201(93)90036-9

Hobara, Y., & Parrot, M. (2005). Ionospheric perturbations linked to a very powerful seismic event. *Journal of Atmospheric and Solar-Terrestrial Physics, 67*(7), 677–685. https://doi.org/10.1016/j.jastp.2005.02.006

Hong, T., Lee, J., Park, S., & Kim, W. (2018). Time-advanced occurrence of moderate-size earthquakes in a stable intraplate region after a megathrust earthquake and their seismic properties. *Scientific Reports, 8*(1), 1–8. https://doi.org/10.1038/s41598-018-31600-5

Humboldt, F. W. H. A. (1845). *Kosmos: Entwurf einer physischen Weltschreibung* (Vol. 6). Gotta.

Ihmlé, P. F., & Jordan, T. H. (1994). Teleseismic search for slow precursors to large earthquakes. *Science, 266*(5190), 1547–1551. https://doi.org/1547

Jellinek, A. M., & Manga, M. (2004). Links between long-lived hot spots, mantle plumes, and plate tectonics. *Reviews of Geophysics, 42*(3). https://doi.org/10.1029/2003RG000144

Jenkins, S. F., Biass, S., Williams, G. T., Hayes, J. L., Tennant, E., Yang, Q., Burgos, V., Meredith, E. S., Lerner, G. A., Syarifuddin, M., & Verolino, A. (2022). Evaluating and ranking Southeast Asia's exposure to explosive volcanic hazards. *Natural Hazards and Earth System Sciences, 22*(4), 1233–1265, https://doi.org/10.5194/nhess-22-1233-2022

Johnson, R. W. (1989). *Intraplate volcanism: In eastern Australia and New Zealand.* Cambridge University Press.

Kanamori, H. (2004). The diversity of the physics of earthquakes. *Proceedings of the Japan Academy. Series B, Physical and Biological Sciences, 80*(7), 297–316. https://www.ncbi.nlm.nih.gov/pmc/articles/PMC8147668/

Keilis-Borok, V. I. (1996). Intermediate-term earthquake prediction. *Proceedings of the National Academy of Sciences, 93*(9), 3748–3755.

Kelman, I., & Glantz, M. H. (2014). Early warning systems defined. In Z. Zommers & A. Singh (Eds.), *Reducing disaster: Early warning systems for climate change* (pp. 89–108). Springer.

Khilyuk, L. F., Robertson, J. O., Endres, B., & Chilingarian, G. V. (2000). *Gas migration: Events preceding earthquakes.* Elsevier.

Koizumi, N., Tsukuda, E., Kamigaichi, O., Matsumoto, N., Takahashi, M., & Sato, T. (1999). Preseismic changes in groundwater level and volumetric strain associated with earthquake swarms off the east coast of the Izu Peninsula, Japan. *Geophysical Research Letters, 26*(23), 3509–3512. https://doi.org/10.1029/1999GL005381

Langbein, J., Borcherdt, R., Dreger, D., Fletcher, J., Hardebeck, J. L., Hellweg, M., et al. (2005). Preliminary Report on the 28 September 2004, M 6.0 Parkfield, California Earthquake. *Seismological Res. Lett., 76*(1), 10–26. doi:10.1785/gssrl.76.1.10

Lay, T., Williams, Q., & Garnero, E. J. (1998). The core–mantle boundary layer and deep Earth dynamics. *Nature, 392*(6675), 461–468. https://doi.org/10.1038/33083

Lewis, J. (2003). Housing construction in earthquake-prone places: Perspectives, priorities and projections for development. *Australian Journal of Emergency Management, 18*(2), 35–44. https://knowledge.aidr.org.au/media/3673/ajem-18-02-04.pdf

Liu, J. Y. (2009). Earthquake precursors observed in the ionospheric F-region. In M. Hayakawa (Ed.), *Electromagnetic phenomena associated with earthquakes* (pp. 187–204). Transworld Research Network.

Liu, J. Y., Chuo, Y. J., Shan, S. J., Tsai, Y. B., Chen, Y. I., Pulinets, S. A., & Yu, S. B. (2004). Pre-earthquake ionospheric anomalies registered by continuous GPS TEC measurements. *Annals of Geophysics, 22*(5), 1585–1593. https://doi.org/10.5194/angeo-22-1585-2004

Liu, W., Li, L., Chen, L., Wen, M., Wang, J., Yuan, L., Liu, Y., & Li, H. (2020). Testing a comprehensive volcanic risk assessment of Tenerife by volcanic hazard simulations and social vulnerability analysis. *ISPRS International Journal of Geo-Information, 9*(4), 273. https://doi.org/10.3390/ijgi9040273

Lowenstern, J. B., Smith, R. B., & Hill, D. P. (2006). Monitoring super-volcanoes: Geophysical and geochemical signals at Yellowstone and other large caldera systems. *Philosophical Transactions of the Royal Society A: Mathematical, Physical and Engineering Sciences, 364*(1845), 2055–2072. http://doi.org/10.1098/rsta.2006.1813

Martinelli, G. (1998). Earthquakes, prediction. In G. A. Good (Ed.), *Sciences of the earth, an encyclopedia of events, people, and phenomena* (pp. 192–196). Garland Publishing.

McCaffrey, R. (2021). Earthquakes and crustal deformation. In H. K. Gupta (Ed.), *Encyclopedia of solid earth geophysics. Encyclopedia of earth sciences series* (pp. 254–262). Springer. https://doi.org.pbidi.unam.mx:2443/10.1007/978-3-030-58631-7_2

McDonough, W. F., & Sun, S. S. (1995). The composition of the earth. *Chemical Geology, 120*(3–4), 223–253. https://doi.org/10.1016/0009-2541(94)00140-4

Mercalli, G. (1902). *Sulle modificazioni proposte alla scala sismica De Rossi-Forel.* Società tipografica modenese.

Mercer, J., & Kelman, I. (2010). Living alongside a volcano in Baliau, Papua New Guinea. *Disaster Prevention and Management, 19*(4), 412–422. https://doi.org/10.1108/09653561011070349

Mignan, A. (2008). The stress accumulation model: Accelerating moment release and seismic hazard. *Advances in Geophysics, 49*, 67–201. https://doi.org/10.1016/S0065-2687(07)49002-1

Molchanov, O. A., & Hayakawa, M. (2008). *Seismo-electromagnetics and related phenomena: History and latest results.* Terrapub.

Németh, K. (2010). Monogenetic volcanic fields: Origin, sedimentary record, and relationship with polygenetic volcanism. In E. Cañón-Tapia & A. Szakács (Eds.), *What is a volcano?*. Geological Society of America Special Paper 470. https://doi.org/10.1130/2010.2470(01)

Newhall, C. G., & Self, S. (1982). The volcanic explosivity index (VEI) an estimate of explosive magnitude for historical volcanism. *Journal of Geophysical Research: Oceans, 87*(C2), 1231–1238. https://doi.org/10.1029/JC087iC02p01231

Niu, F., Silver, P. G., Daley, T. M., Cheng, X., & Majer, E. L. (2008). Preseismic velocity changes observed from active source monitoring at the Parkfield SAFOD drill site. *Nature, 454*(7201), 204–208. https://doi.org/10.1038/nature07111v

Oppenheimer, C. (2002). Limited global change due to the largest known quaternary eruption, Toba ≈74 kyr BP? *Quaternary Science Reviews, 21*(14–15), 1593–1609. https://doi.org/10.1016/S0277-3791(01)00154-8

Proud, S. R., Prata, A. T., & Schmauß, S. (2022). The January 2022 eruption of Hunga Tonga-Hunga Ha'apai volcano reached the mesosphere. *Science, 378*(6619), 554–557. https://doi.org/abo4076

Pulinets, S., Legen'ka, A., Gaivoronskaya, T., & Depuev, V. (2003). Main phenomenological features of ionospheric precursors of strong earthquakes. *Journal of Atmospheric and Solar-Terrestrial Physics, 65*(16–18), 1337–1347. https://doi.org/10.1016/j.jastp.2003.07.011

Pulinets, S., & Ouzounov, D. (2018). *The possibility of earthquake forecasting: Learning from nature.* IOP publishing.

Rapoport, Y., Grimalsky, V., Krankowski, A., Pulinets, S., Fedorenko, A., & Petrishchevskii, S. (2020). Algorithm for modeling electromagnetic channel of seismo-ionospheric coupling (SIC) and the variations in the electron concentration. *Acta Geophysica, 68*(1), 253–278. https://doi.org/10.1007/s11600-019-00385-0

Richter, C. F. (1935). An instrumental earthquake magnitude scale. *Bulletin of the Seismological Society of America, 25*(1), 1–32.

Rodger, C. J., Clilverd, M. A., & Thomson, N. R. (1999). Modeling of subionospheric VLF signal perturbations associated with earthquakes. *Radio Science, 34*(5), 1177–1185. https://doi.org/10.1029/1999RS900061

Rostain, S., Dorison, A., De Saulieu, G., Prümers, H., Le Pennec, J. L., Mejía Mejía, F., Freire, A. M., Pagán-Jiménez, J. R., & Descola, P. (2024). Two thousand years of garden urbanism in the Upper Amazon. *Science, 383*(6679), 183–189. https://doi:10.1126/science.adi6317

Scandone, R., Bartolini, S., & Martí, J. (2016). A scale for ranking volcanoes by risk. *Bulletin of Volcanology, 78*(1), 1–8. https://doi.org/10.1007/s00445-015-0995-y

Scholz, C. H. (2019). *The mechanics of earthquakes and faulting* (3rd ed.). Cambridge University Press. https://doi.org/10.1017/9781316681473

Scott, W. E. (1989). Volcanic and related hazards. In R. I. Tilling (Ed.), *Volcanic hazards* (pp. 9–23). American Geophisical Union. https://doi.org/10.1029/SC001p0009

Self, S. (1992). Krakatau revisited: The course of events and interpretation of the 1883 eruption. *GeoJournal, 28*, 109–121. https://doi.org/10.1007/BF00177223

Self, S. (2005). Effects of volcanic eruptions on the atmosphere and climate. In J. Marti & G. G. J. Ernst (Eds.), *Volcanoes and the environment* (pp. 152–174). Cambridge University Press.

Self, S. (2006). The effects and consequences of very large explosive volcanic eruptions. *Philosophical Transactions of the Royal Society A: Mathematical, Physical and Engineering Sciences, 364*(1845), 2073–2097. http://doi.org/10.1098/rsta.2006.1814

Shearer, P. M. (2019). *Introduction to seismology*. Cambridge University Press.

Sigurdsson, H., Houghton, B., McNutt, S., Rymer, H., & Stix, J. (Eds.). (2015). *The encyclopedia of volcanoes*. Elsevier.

Sorokin, V. M., Chmyrev, V. M., & Hayakawa, M. (2020). A review on electrodynamic influence of atmospheric processes to the ionosphere. *Open Journal of Earthquake Research, 9*(2), 113–141. https://doi.org/10.4236/ojer.2020.92008

Sparks, R. S. J., Self, S., Grattan, J., Oppenheimer, C., Pyle, D., & Rymer, H. (2005). *Super-eruptions: Global effects and future threats* (pp. 1–24). Geological Society, Report of a Geological Society of London Working Group. https://www.geolsoc.org.uk/~/media/shared/documents/education%20and%20careers/Super_eruptions.pdf?la

Suárez, G. (2022). The seismic early warning system of Mexico (SASMEX): A retrospective view and future challenges. *Frontiers in Earth Science, 10*, 827236. https://doi.org/10.3389/feart.2022.827236

Szakács, A. (1994). Redefining active volcanoes: A discussion. *Bulletin of Volcanology, 56*(5), 321–325. https://doi.org/10.1007/BF00326458

Taha, G., Loughman, R., Colarco, P. R., Zhu, T., Thomason, L. W., & Jaross, G. (2022). Tracking the 2022 Hunga Tonga-Hunga Ha'apai aerosol cloud in the upper and middle stratosphere using space-based observations. *Geophysical Research Letters, 49*(19). https://doi.org/10.1029/2022GL100091

Tilling, R. I. (2005). Volcano hazards. In J. Marti & G. G. J. Ernst (Eds.), *Volcanoes and the environment* (pp. 55–89). Cambridge University Press.

Tramutoli, V., Cuomo, V., Filizzola, C., Pergola, N., & Pietrapertosa, C. (2005). Assessing the potential of thermal infrared satellite surveys for monitoring seismically active areas: The case of Kocaeli (İzmit) earthquake, August 17, 1999. *Remote Sensing of Environment, 96*(3–4), 409–426. https://doi.org/10.1016/j.rse.2005.04.006

USGS. (n.d.). *Yellowstone volcano observatory*. http://volcanoes.usgs.gov/yvo/index.html

Uyeda, S., Nagao, T., & Kamogawa, M. (2009). Short-term earthquake prediction: Current status of seismo-electromagnetics. *Tectonophysics, 470*(3–4), 205–213. https://doi.org/10.1016/j.tecto.2008.07.019

Vogelmann, A. M., Ackerman, T. P., & Turco, R. P. (1992). Enhancements in biologically effective ultraviolet radiation following volcanic eruptions. *Nature, 359*(6390), 47–49. https://doi.org/10.1038/359047a0

Walker, G. P. (1999). Volcanic rift zones and their intrusion swarms. *Journal of Volcanology and Geothermal Research, 94*(1–4), 21–34. https://doi.org/10.1016/S0377-0273(99)00096-7

Wegener, A. (1912). Die entstehung der kontinente. *Geologische Rundschau, 3*(4), 276–292.

White, J. D., McPhie, J., & Soule, S. A. (2015a). Submarine lavas and hyaloclastite. In *The encyclopedia of volcanoes* (2nd ed., pp. 363–375). https://doi.org/10.1016/B978-0-12-385938-9.00019-5

White, J. D., Schipper, C. I., & Kano, K. (2015b). Submarine explosive eruptions. In *The encyclopedia of volcanoes* (2nd ed., pp. 553–569). https://doi.org/10.1016/B978-0-12-385938-9.00031-6

Woith, H., Petersen, G. M., Hainzl, S., & Dahm, T. (2018). Can animals predict earthquakes? *Bulletin of the Seismological Society of America, 108*(3A), 1031–1045. https://doi.org/10.1785/0120170313

Wood, H. O., & Neumann, F. (1931). Modified Mercalli intensity scale of 1931. *Bulletin of the Seismological Society of America, 21*(4), 277–283.

Yanovskaya, T. B. (2008). *Basics of seismology*. Publishing House University.

Zhan, Z. (2020). Mechanisms and implications of deep earthquakes. *Annual Review of Earth and Planetary Sciences, 48*, 147–174. https://doi.org/10.1146/annurev-earth-053018-060314

Zhao, W., Sun, C., & Guo, Z. (2022). Reawaking of Tonga volcano. *The Innovation, 3*(2), 100218. https://doi.org/10.1016/j.xinn.2022.100218

Chapter Seven
Living with Tsunamis and Landslides

Tsunamis

Surimi, umami . . . what was the word? . . . aazhi peralai, ië beuna, alôn buluël, among, emong, zeebeben, maremoto, 津波.

Have you ever heard of the famous painting *The Great Wave of Kanagawa*? This woodblock print was created by a Japanese ukiyo-e artist, Hokusai, during the Edo period of Japanese history (1603–1876). The painting depicts three boats descending through a turbulent sea, with a massive spiral wave at the center and Mount Fuji visible in the distance. The waves shown in the painting are known as *okinami* (沖波), which are impressive offshore waves and should not be mistaken for tsunamis (津波), a term that is translated as a "wave in the harbor". However, the artwork's grandeur may evoke the possible impact of a colossal wave caused by an earthquake, volcanic eruption, slide, or strike from a space object.

Gupta and Gahalaut (2013) found that very few languages have a native word for tsunami. Tamil Nadu, an east coast province of India, uses the word *aazhi peralai* (destruction, big waves). Aceh, Sumatra, Indonesia, uses *ië beuna* or *alôn buluël*. Simeulue Island, off the western coast of Sumatra in Indonesia, uses *smong* and *emong* in Defayan and Sigulai languages, respectively. Other terms used include high-tide wave, seismic sea wave, seaquake, and the antique European terms *zeebeben* and *maremoto*, a word in Spanish (Levin et al., 2016).

What Causes Tsunamis?

Tsunamis are gravity-driven, long, shallow water waves that can happen in oceans or vast lakes. They differ from common sea waves, as they are frequently generated by sudden vertical ocean floor movement, while sea waves evolve from persistent winds or tidal effects (although meteotsunamis are discussed later). Their occurrence is due to an

DOI: 10.4324/9781003292814-7

unexpected shift in the water column caused by multiple factors, like earthquakes, volcanic eruptions, cosmic impacts, and submarine or subaerial landslides (Röbke & Vött, 2017).

Various factors, such as explosions on volcanic islands, eruptions of underwater volcanoes, gravitational collapses of volcanic flanks, and pyroclastic flows in water, can cause tsunamis. One of the most significant tsunamis caused by a volcano occurred in August 1883 when Krakatau erupted in the Indonesian archipelago (Self & Rampino, 1981). The aftermath of the explosions resulted in a series of ash clouds and small tsunamis, followed by a more severe tsunami from the collapse of mountain Danan. The third explosion created a tsunami that was 42 m high and reached 5 km inland. This tsunami caused significant destruction, killing at least 36,000 people and destroying about 300 villages. The tsunami was noticed worldwide, with waves recorded by many mareographs in the Indian Ocean, Pacific, and Atlantic (Levin & Nosov, 2016).

A volcano in the Tofua Volcanic Arc, Hunga Tonga-Hunga Ha'apai, erupted on 14 January 2022, causing tsunamis and global sea-level changes (Pararas-Carayannis, 2022). Data from 121 tide gauge stations in 28 countries showed that tsunami wave height varied depending on location, wave velocity, and coastal topography. For example, the wave height in Japan was 1.20 m, while Costa Rica's was only 0.11 m. The wave height also varied in other countries, such as Chile, Mexico, and the United States of America (Jonathan et al., 2023).

Various factors, including collisions of cosmic bodies with oceans and seas, can cause tsunamis. As a result of these collisions, exceptionally large waves can be created, as Wünnemann and Weiss (2015) reported. Such impacts causing tsunami deposits are a significant concern as future strikes remain possible (Bryant & Nott, 2001). Predicting impacts and identifying past deposits to understand better the processes are complex tasks, requiring further research to fully comprehend the implications.

Small asteroids cause minimal damage upon entering Earth's atmosphere, but those ranging from 200 m to 1 km can trigger catastrophic tsunamis. A 400 m wide asteroid can create tsunamis over 100 m in height around an ocean, and a 5 km asteroid impact would submerge the entire upper East Coast of the United States (Hills & Mader, 1997) (see Chapter 8).

Levin and Nosov (2016) identified several causes of landslide-induced tsunamis, including underwater landslides, mudflows, steep coasts, rocks, icebergs, or collapsing buildings. These tsunamis can be highly destructive (see Box 7.1) and are common in narrow straits, fjords, closed gulfs, bays, some coastal regions, and large rivers.

BOX 7.1 The Tsunami of Tōhoku, Japan, 2011: More than an Earthquake-Induced Tsunami

The devastating tsunami during the 2011 Great East Japan Earthquake (Figure 7.1) had two causes: vertical seafloor displacement resulting from the earthquake and a submarine landslide. Researchers, led by Tappin et al. (2014), discovered

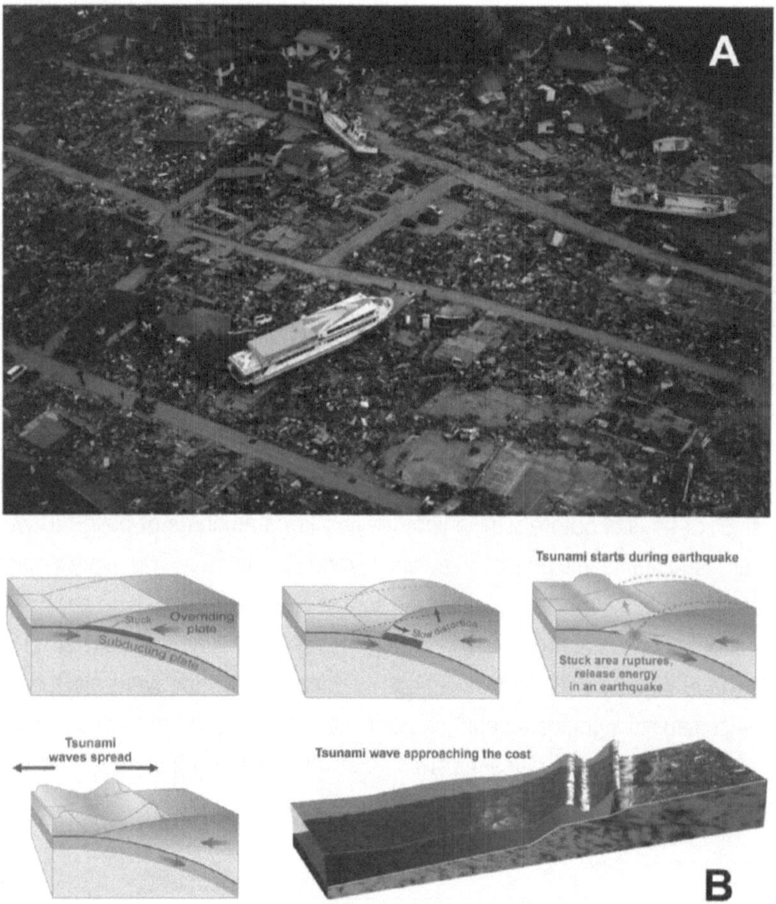

FIGURE 7.1 A. The impact was felt in different areas, including Miyako, Iwate Prefecture, Japan, where a large ferry boat rests inland amidst destroyed houses (20 March 2011)
Source: U.S. Marine Corps photo by Lance Cpl. Garry Welch, Wikimedia Commons

B. Tsunamis are primarily induced by earthquakes

Source: U.S. Geological Survey Circular 1187 https://pubs.usgs.gov/circ/c1187/ and http://de.wikipedia.org/wiki/Benutzer:Veitmueller

that this combination was responsible for the massive 40 m high waves that hit the Sanriku district, as well as the frequency and timing of the tsunami waves recorded by three nearshore buoys off Sanriku. The study also found that the earthquake's horizontal acceleration caused several meters of vertical seafloor movement, enough to trigger the submarine landslide. These findings are crucial in estimating tsunami risks in similar tectonic environments, as multiple hazards can occur simultaneously.

Source: Tappin et al. (2014)

BOX 7.2 Meteotsunamis

Meteotsunamis, a variant of tsunamis, are generated by meteorological factors, including squalls, thunderstorms, frontal passages, and atmospheric gravity waves, and have the potential to cause significant damage. Even small perturbations in sea level of a few centimeters can be amplified through multi-resonant phenomena, causing devastating consequences. An instance of such an event occurred on 21 June 1978 in Vela Luka, Croatia. The first wave flooded basements, broke house walls, and pushed boats ashore. A second wave reached almost the first floor of houses before weakening, leaving large amounts of olive oil, wine, furniture, and household goods mixed and spread over the bay. The event's resemblance to a tsunami was obvious, but the source was unclear since no earthquake was recorded. The flood damaged hotels, a fish factory, and homes, but no one died. The electricity was shut down, and national authorities were alerted. The flood also caused significant pollution from septic tanks, and the local community feared the possibility of tsunamis. This event inspired two local poets to write about it.

Sources: Pattiaratchi & Wijeratne (2015); Vučetić et al. (2009)

Approximately 90% of tsunamis are produced by shallow-focus (focal depth < 70 km) tectonic earthquakes with a large magnitude (> M 6.5), causing a vertical displacement in the ocean floor, resulting in a displacement of the overlying water column. Earthquake-generated tsunamis are primarily caused by reverse and normal fault-dominated earthquakes. Tsunamis mostly happen in the Pacific Ocean, especially in the subduction zones around the Ring of Fire zone. The Java-Sumatra zone is also a tsunamigenic area. Tsunamis have occurred from earthquakes in places like the Makaran region, the Mediterranean Sea, and near Portugal (Gupta & Gahalaut, 2013).

Dynamics of Tsunamis: Focusing on Fundamental Notions

Tsunamis are different from regular ocean waves. They are long-wavelength waves with small heights in deep oceans but can increase in shallow water. They can travel long distances with minimal energy loss and form circular rings with ray paths resembling spokes on a wheel (Gupta & Gahalaut, 2013) when propagating from their source. Teletsunamis, also called remote tsunamis, can travel thousands of kilometers before hitting distant shores (Röbke & Vött, 2017). These tsunamis are typically caused by seismic activity that occurs more than 1,000 km away from the affected coastline. This distance can also translate to more than three hours of travel time for the tsunami. Floor shape, distance to the source and island chains, and wave height onshore, depending on energy and shoreline shape, affect tsunami speed.

The ocean may recede before a tsunami, revealing previously water-covered areas. Some people who do not know the danger might stay near the shore or walk onto the empty sea floor to look around or collect fish. They can drown when the tsunami crest comes. Tsunamis pose a significant hazard, causing floods, damaging buildings and bridges, eroding homes, and destroying coastal infrastructure, leading to secondary effects, such as fires, pollution, and epidemics (Klug, 1986; Gupta & Gahalaut, 2013; Levin & Nosov, 2016) (see Box 7.3).

BOX 7.3 Tsunami Family Saved by Schoolgirl's Geography Lesson

On 26 December 2004, 10-year-old Tilly Smith was on a Thai beach with her family when she noticed something was wrong. She remembered a geography lesson from Andrew Kearney just two weeks before. The Smith family from southeast England celebrated Christmas at Maikhao Beach in Phuket, southern Thailand. A powerful earthquake struck northern Sumatra early in the morning, triggering a dangerous tsunami.

Tilly, who had studied tectonic plates and earthquakes at school, recognized the signs of an earthquake under the sea while her mother noticed the beach getting smaller. She became hysterical and screamed at her parents to leave the beach. Tilly's father took her and her 7-year-old sister back to the hotel pool, while other tourists on the beach were alerted by Tilly's concerns. As Tilly feared, her mum realized something serious was happening, recalled a yacht being tipped over in the bay and the entire sea appearing to come out of the water, and yelled for everyone to run. The Smith family found safety on the third floor of their hotel during the tsunami. They witnessed everything being swept into the swimming pool, including beds, palm trees, and more. They were grateful for avoiding the beach. Afterwards, they encountered individuals from neighboring resorts who had tragically lost all their family members.

Source: Owen (2005)

Measuring Tsunamis and Their Impact

Accurate measurement is crucial for tsunamis. In the past, people used different methods like asking locals, timing when the waves arrive, and measuring how high the waves go. These methods can take a long time to get results. Nowadays, other ways, like using satellites and tide gauges or employing robots with special radar, help survey the aftermath of a tsunami and monitor how the sea level changes in remote areas (Kurkin et al., 2016).

Over time, experts have changed the way they measure the size of tsunamis. The first attempt was the Imamura-Iida scale, which had six points ranging from –1 to 4. Recently, Boschetti and Ioualalen (2021) created a new scale for studying tsunami hazards. This scale looks at the maximum wave currents (CURmax) and the maximum tsunami amplitude (Hmax). It helps experts understand how tsunamis could affect land and sea structures and suggests how to manage and prevent potential problems.

From Tsunami Measurements to Early Warning

Tsunami warning systems were established in the Pacific after devastating events, such as 1946 Aleutian Islands and 1960 Chile. From a merely technological perspective, diverse investigations have suggested that improvements are needed for faster detection, precise evaluation, and continued international cooperation. Likewise, real-time monitoring, advanced methods for earthquake source characterization, and tsunami forecasting are critical, and dedicated warning centers should be continuously supported for coastal safety (Igarashi et al., 2011). Improved understanding of tsunamis has come from 30 years of technological advancements and documentation. Much work remains, especially as the severe consequences of disasters involving tsunamis continue to drive progress (Tappin, 2018) (see Box 7.4).

By definition, early warning systems (EWSs) should not solely focus on technology but also consider the social context. EWSs must address the root causes of disasters, risk perception, and vulnerability and promote adequate risk communication. Legal and ethical frameworks should be in place to create awareness of potential dangers. A transdisciplinary approach is required to understand the underlying causes and drivers of risk, the social and economic context of disasters, and the various dimensions of vulnerability (including exposure). Prioritizing this approach will lead to more effective EWSs and a society that is more aware of risks.

Throughout history, it has been observed that intergenerational warnings are instrumental in averting human casualties during disasters. This practice entails transferring expertise and wisdom from prior generations to current ones, enabling them to prepare and respond effectively to potential disasters. The effectiveness of this approach has

BOX 7.4 The Psychological and Social Impact of Tsunamis

A review of dozens of studies revealed that the Tōhoku tsunami left many people with psychological trauma and mental problems manifesting as anxiety, sleep disorder, depression, and post-traumatic stress disorder (PTSD). The subsequent damage to the Fukushima nuclear facility had similar effects, and the social impact of the workers and nearby affected populations suffered stigma. US$360 billion is one estimate of this disaster's economic impact which destroyed 138,000 buildings. The evacuation of nearly half a million people resulted in the displacement of whole communities. Disruption in manufacturing supply chains had many knock-on effects resulting in a decline in exports and large trade deficits. Recovery efforts were marked by the increasing importance of civil-society-based volunteerism, a long-standing tradition in Japan.

Sources: Harada et al. (2015); Ferris & Solis (2013)

been demonstrated in numerous instances, underscoring the significance of bequeathing knowledge and insights gleaned from past experiences.

Scattered along the vast coastline of Japan, there are numerous sturdy stone slabs thoughtfully placed to serve as essential alerts of impending tsunamis. These markers, called *tsunamihi*, have witnessed the immense power of tsunamis that have been known to wreak havoc on the country's shores with catastrophic consequences. The stone slabs

FIGURE 7.2 Japanese intergenerational tsunami warnings

A. Monuments of the 1792 Unzen earthquake and tsunami in Uki City, Kumamoto Prefecture, Japan. The tsunami reached this stone at a sea level of 20 m

Source: Mizushimasea https://commons.wikimedia.org/wiki/File:Tsunami_Stones_Uki_Kumamoto_1.jpg

B. Print illustrating the tsunami that followed the 7.2 magnitude Meiji Sanriku earthquake of 1896

Source: University of British Columbia Library – Rare Books and Special Collections

are inscribed with phrases such as "If an earthquake comes, beware of tsunamis" and "Remember the calamity of the great tsunamis. Do not build any homes below this point". These words serve as a reminder of the immense destruction that tsunamis can bring and urge people to take necessary precautions to keep themselves and their loved ones safe (Tkalich, 2011) (Figure 7.2).

Along the same line, experiences from Indonesia, Thailand, India, and Oceania have demonstrated the value of various forms of Indigenous knowledge ranging from intangible cultural values and stories to tangible stones or musical instruments. In the case of Indigenous knowledge, knowledge transfer is a personal process based on the principles of direct communication, whether verbal or nonverbal, between master and disciple, parent and child, from neighbor to neighbor, or between leaders and the whole village.

Indigenous knowledge comprises intangible cultural values and tangible objects, which have proven valuable. Personal relationships, such as master and disciple or parent and child, play a crucial role in the transfer of knowledge (Box 7.5). One noted example of this importance can be seen in the Vanuatu Archipelago, where frequent earthquakes are experienced. In 1999, a 7.5 Mw earthquake triggered a tsunami that destroyed Baie Martelli village. Although the tsunami arrived at night, almost all 300 residents survived. This was made possible by the quick actions of a group of individuals who immediately recognized the receding sea level, took necessary precautions, and promptly fled to the safety of the nearby hills (Mikulecký et al., 2023). This underscores the importance of Indigenous knowledge and its critical role in preserving and protecting communities.

BOX 7.5 The Oral Histories of the People of Simeulue: The Ocean Coming onto the Land

Simeulue, the northernmost island in the chain off the west coast of Sumatra, has faced numerous challenges. However, the local people have shown remarkable capacity and progress, learning from past experiences. In 1907, a MW $\approx$ 8.2 earthquake caused a devastating tsunami that destroyed the island and much of the Sumatran coast. The local people called it a "smong", which means "the ocean coming onto the land", and the term remains in use today. In 2004, an earthquake of 9.2 Mw struck less than 50 km from Simeulue, triggering a tsunami that affected the island. Nonetheless, thanks to their oral histories and quick reactions, 44 deaths were reported despite the population growth.

The earthquake happened, my grandmother said. Her skin was wrinkled but her hair was still black, and her teeth were good. She said "If earthquake happens run, don't bring anything. After the earthquake take the rice, the water, clothes, trousers and check the sea, if the water recedes 'RUN FAR' ", she said – shaking her finger.

(Sutton et al., 2020, p. 6)

Due to their understanding of tsunamis and how to survive them, they were able to transform their story and successfully evacuate the coast of the island. The low death toll is a testament to their strength and perseverance. A combination of high-tech tsunami warning systems, such as a full network of satellite-linked "tsunameters", and local grassroots education, is necessary.

Despite this favorable experience, the response of the different ethnic groups of Aceh to the 2004 earthquake and tsunami was influenced by the presence of a disaster subculture in some communities. Cultural, economic, and political constraints also played a significant role in shaping behavior. Religion and the armed conflict were major constraints.

Sources: McAdoo et al. (2006); Sutton et al. (2020); Gaillard et al. (2008)

Supporting traditional knowledges in Indigenous communities helps to improve the understanding and preparedness for tsunamis. Transferring this knowledge should be respectful of individual communities, balancing knowledge availability with proprietary aspects. Different viewpoints emerge regarding cultural (mis)appropriation and paying for commercial data products while capturing, preserving, and making freely available Indigenous, local, and vernacular knowledges. Various knowledges together, including but not limited to modern technologies, aid in predicting and mitigating tsunami risk.

Landslides: Causality and Classification of Landslides

Landslides are various mass movement processes involving materials that form slopes moving downward and outward, driven by gravity (Brunsden, 1979), although they can bounce around, being deflected upwards and converging. Landslides are exogenic geomorphological processes that modify landforms on the Earth's surface.

The slope can fail when the force pushing down on a slope exceeds the soil's shear strength. Steeper slopes are more susceptible to failure due to increased shear stress. Shear strength depends on soil particles' cohesion and the friction between them. Soils with stronger shear strength are more deformation-resistant (Terzaghi,1950; Skempton & Hutchinson, 1969).

External and internal factors can cause the instability of slopes. Heavy rainfall, earthquakes, and human activities can increase the forces acting on a slope. At the same time, the type of materials, their orientation, and the presence of discontinuities can reduce the resistance of materials. These factors can work together to concentrate driving forces and decrease resistant forces, ultimately leading to slope instability (Cruden & Varnes, 1996).

Therefore, the stability of a slope relies on resisting forces exceeding driving forces. Driving forces consist of the weight of soil or rock and any structures, while resisting forces come from the inherent strength and durability of the slope. Understanding these factors is crucial for designing infrastructure, buildings, and land use in areas vulnerable to mass movement processes. Unfortunately, these decisions are sometimes made either in the absence of regulations, despite their contravening regulations, or sometimes with the complicity of local officials (see Box 7.6 and Figure 7.3). For instance, officials in Uganda have so much interest in promoting the expansion of coffee growing that they encourage farmers to cut forest vegetation on the steep slopes of Mount Elgon whilst ignoring the known risk of landslides and the long record of their occurrence.

BOX 7.6 Landslides in Petrópolis, Brazil

Petrópolis is located within the Atlantic Forest biome, although only 13% of the original forest remains (Fundação SOS Mata Atlântica, 2018) due to tourism growth, real estate speculation, and expansion into landslide-prone areas near the historical center (Marchezini & Wisner, 2017). The population has increased rapidly, reaching 75,000 in 1940 and 255,000 in 1990, with mass movements occurring during the 1980s, even though rainfall values were lower than in previous decades (Guerra et al., 2007).

Due to its geological and geomorphological character, the city experiences frequent landslides. Over 26% of the area comprises 37 degrees or more slopes, with soil-rock contact, water flow convergence, and failed and fractured rocks contributing to the instability (Guerra et al., 2007). The densely populated areas, particularly the slopes, are most affected.

In January 2011, an area of 350 square km experienced over 750 landslides, averaging at least two per square kilometer. The most common types of landslides were debris flows and mudflows. On 12 January 2011, there was a total rainfall of 240 mm, with a peak of 61 mm in one hour. Between 1 January and 15 January, there was a total of 573 mm of rainfall (Dourado et al., 2012). Of the people living in areas prone to floods and landslides, 26% live in subnormal agglomerates (slums). Among them, 54% have no water supply, and 14% have inadequate sanitation services (Assis Dias et al., 2018).

Sources: Assis Dias et al. (2018); Dourado et al. (2012); Fundação SOS Mata Atlântica (2018); Guerra et al. (2007); Marchezini & Wisner (2017)

FIGURE 7.3 Slums of Peru

Source: photo by Alex Proimos, Wikimedia Commons

Factors influencing and triggering landslides can be both natural processes and human activities (Table 7.1). Natural triggers include heavy rain, snowmelt, earthquakes, volcanic eruptions, permafrost thawing, flooding, and drought-induced cracking (Zeng et al., 2020). Human-induced triggers include excavation at the base of a slope, water leakage, irrigation, drainage diversion, blasting or other vibrations, site loading, and rapid changes in reservoir levels (Geertsema & Alcántara-Ayala, 2022).

Mass movement processes can be classified in various ways, and their study can be complex and interdisciplinary. Typically, these classifications are based on several factors, such as the physical attributes of the movement, which include the shape of the rupture surface and the deposit area. Other aspects considered are the type and speed of the movement, as well as the size and type of materials involved. Additionally, the age of the

TABLE 7.1 Main landslide preconditioning factors

Factor	Result
Preconditioning (Longer term)	
Chemical and physical weathering	Weakens rock masses
Leaching of salt	Development of sensitive soils
Stream incision	Deepens valleys and steepens slopes
Glacial erosion	Results in valley deepening – imparts stresses on slopes
Glacial retreat	Glacial debuttressing leads to slope adjustments and sometimes catastrophic failure
Tectonics	Strain, stress, steepening, tilting strata, etc.
Wildfire	Loss of root strength; hydrophobicity
Soil development	Formation of hardpans
Volcanoes	Flank collapse
Natural causes (Shorter term)	
High stream flow	Bank erosion triggers
Rainfall	High pore water pressures initiating slides and flows
Snowmelt	
Permafrost thaw	Active layer detachments, bedrock failure, thaw landslides
Earthquake	Rock landslides, liquefaction, soil landslides
Volcanoes	Lahars
Anthropogenic causes	
Blasting, pile driving	Liquefaction triggered landslides
Deforestation	Loss of root strength and evapotranspiration-related landslides
Cut-and-fill road construction	
Improper drainage along roads	Road-related landslides
Site loading	Landslides from undrained loading
Dam construction	Reduce replenishment of beaches – coastal erosion

Source: Alcántara-Ayala & Geertsema (2022)

movement, and its level of activity, may be considered when classifying these types of processes (Skempton & Hutchinson, 1969; Varnes, 1978; Zaruba & Mencl, 2014).

One of the most frequently employed means of categorizing landslides is through their respective movement mechanisms. This approach allows for a more detailed understanding of the specific ways these events occur and their potential impact. The mechanics of landslides are multifaceted and can manifest in several forms, including falling, toppling, sliding, spreading, flowing, or slope deformation. Furthermore, it is not uncommon for landslides to arise from a combination of two or more types of movement, known as complex movements. Several authors have identified different types of landslides based on the current understanding of landslide dynamics. These primary factors are widely used to characterize landslides in both research and practice (see Table 7.2).

TABLE 7.2 Classification of landslide types and main characteristics

Type of movement	Main characteristics	Rock	Soil
Fall	Falls involve the detachment of soil, rock, and ice fragments from their original location and falling to the ground. These fragments may then roll or bounce either by themselves or in clusters.	*Rock/ice* fall[a]	*Boulder/debris/silt* fall[a]
Topple	Toppling is when a rock column or plate rotates forward and overturns due to steeply dipping joints. It can happen on any scale. Rock flexural toppling involves the forward rotation and bending of weak and fissile rock masses with no well-defined basal joints.	Rock block topple[a] Rock flexural topple	*Gravel/sand/silt* topple[a]
Slide	Slides occur when soil, rock, or debris moves down a slope. Transverse cracks appear, indicating the onset of a landslide. The shape of the rupture surface determines the type of landslide: rotational, translational, or planar.	Rock rotational slide Rock planar slide[a] Rock wedge slide[a] Rock compound slide Rock irregular slide[a]	*Clay/silt* rotational slide *Clay/silt* planar slide *Gravel/sand/debris* slide[a] *Clay/silt* compound slide
Spread	Spread is when soil or rock moves sideways over soft material caused by high water pressure or earthquakes. The movement is triggered by high water pressure caused by rainfall or snowmelt, earthquakes, or high-water content in clays.	Rock slope spread	*Sand/silt* liquefaction spread[a] Sensitive clay spread[a]

(*Continued*)

TABLE 7.2 (Continued)

Type of movement	Main characteristics	Rock	Soil
Flow	Flow is when soil or slopes act like a liquid, especially when composed of silt and clay. Particles move at different speeds during a flow, which can happen in any material. It is often caused by previous mass movements that create excess pressure. Soil flows can also occur in dry sands, which are rare but dangerous due to high velocity.	*Rock/ice* avalanche[a]	*Sand/silt/debris* dry flow *Sand/silt/debris* flowslide[a] Sensitive clay flowslide[a] Debris flow[a] Mud flow[a] Debris flood Debris avalanche[a] Earthflow Peat flow
Slope deformation	Mountain slopes deform due to gravity, resulting in scarps, benches, cracks, trenches, and bulges. The process does not always involve a defined rupture surface, and movement rates can be slow or hard to measure.	Mountain slope deformation Rock slope deformation	Soil slope deformation Soil creep Solifluction

Source: Hungr et al. (2014)

Sources: Varnes (1978); Cruden & Varnes (1996); Leroueil et al. (1996); Highland & Bobrowsky (2008); Clague & Stead (2012); Hungr et al. (2014)

The words in italics are placeholders (use only one type), a movement type that usually reaches extremely rapid velocities, as defined by Cruden and Varnes (1996). The other landslide types are often extremely slow to very rapid (but not always).

In essence, landslides can be viewed as physical systems that undergo distinct stages of development: pre-failure deformations, failure, and post-failure movement of materials that can cause harm to people and property (Terzaghi,1950; Skempton & Hutchinson, 1969; Leroueil et al., 1996).

Evaluating landslides requires diverse research techniques, including aerial and field reconnaissance. Creating detailed landslide maps at various scales, such as inventory, susceptibility, hazard, and risk, provides insight into their occurrence. GIS, remote sensing, and real-time monitoring simplify landslide analysis and improve our understanding of movement rates and material properties. Evaluating landslides also involves understanding contributing factors and triggers (see Highland & Bobrowsky, 2008).

Landslide Impact on Society

Landslides can be involved in significant damage and loss of life. Monitoring and mitigating their effects through soil stabilization, drainage systems, and structural reinforcement is essential. Additionally, avoiding human settlements in landslide-prone areas is fundamental to avoiding disasters (Figure 7.4).

Climate change seriously impacts mountain regions, causing rising temperatures, changes in weather patterns, and loss of glaciers (see Chapter 5). If global warming reaches 1.5°C, many small glaciers will lose most of their mass, impacting biodiversity and human well-being. Rounce et al. (2023) have suggested that the impact of global warming on mountain glaciers is a cause for concern, as it could result in a loss of up to 50% of their mass by the year 2100. This could lead to a greater contribution to the rise of sea levels than what was previously estimated. Nonetheless, it must be recognized that human agency creates vulnerabilities, exposes vulnerable people to hazards, and shapes and modifies them.

FIGURE 7.4 Human settlements are often developed in areas highly susceptible to landslides. Aerial view (left image) and close-up (right image) of Teziutlán, Puebla, México, a village historically affected by landslides

Source: photo by Ricardo J. Garnica-Peña

Landslide Hazard Mitigation Measures

Reducing the impact of landslides is complex and requires a tailored approach. Measures include stabilizing soil, improving drainage, and using bioengineering techniques. Structural measures like debris fences can also help. Non-structural methods involve raising public awareness, co-producing knowledge, enforcing land use laws, developing policies and regulations, and promoting social cohesion (Alcántara-Ayala & Geertsema, 2022) (Table 7.3).

TABLE 7.3 Landslide structural and non-structural measures

STRUCTURAL	
Strategy domain	Landslide type – Actions
Deep slides	
Loading the toe	Place a rock apron on a lower slope as a counterweight, especially for rotational slides.
Unloading the upper slope	Remove weight from the upper slope. Employ setbacks to prevent loading or construction within a certain distance from the slope break.
Surface drainage	Reestablish natural drainage where old road networks exist above unstable slopes.
Internal drainage	Install horizontal drains. Install vertical drains.
Shallow slides	
Bioengineering	Hydroseeding Installing horizontal root wattles
Debris fences	Soil anchors Tunnels Shedding structures
Rockfall	
	Rock bolts for rockslides Wire mesh to control rockfall Rock scaling to control rockfall Debris fences to protect against rockfall Forested slopes between cliffs and elements at risk
Debris floods and flows	
Upper slope	Baffles Check dams
Lower slope	Lateral berms Deflection berms Catch basins

(Continued)

TABLE 7.3 (Continued)

NON-STRUCTURAL	
Strategy domain	Cornerstones – Actions
Knowledge management efforts	
Public awareness raising	Dissemination of information to the public on a regular basis Publication of warning signs Real-time weather data Development of landslide scenarios Establishment and dissemination of landslide early warning systems Elaboration of landslide disaster risk communication strategies
Education	Understanding landslide disaster risk through education strategic plans Developing education programs to understand landslide dynamics, landslide symptoms in urban and rural contexts, and landslide warning systems from a nontechnical perspective Implementation of initiatives for capacity building
Training	Capacity building directed toward understanding/undertaking in the field: (a) Landslide warning systems (b) Landslide evacuation drills (c) Warning and evacuation training (collective and self-evacuation procedures) Training of local officials in charge of disaster prevention Training of local communities Training of all community to understand and apply contingency plans
Scientific knowledge	Landslide zoning and warning Landslide inventories and mapping Landslide hazard assessment Landslide risk assessment Vulnerability analysis Exposure mapping Landslide disaster risk scenarios Integrated landslide disaster risk research
Traditional knowledge	Landslide inventories and mapping Identification of areas susceptible to landslides Recollection of stories and experiences about old landslide
Co-production of knowledge	Landslide mapping Landslide risk analysis Identification of feasible actions to manage disaster risk
Agreement	
Land use and environmental laws, bylaws, regulations, and controls	Generating/updating/enacting environmental laws and regulations Generating/updating/enacting land use laws and regulations Land use integrated management Zoning ordinances for limiting types of land use Private land use restrictions Regulation of existing human settlement on hillslopes Preventing development in areas susceptible to landslides Restricting the use of areas susceptible to landsliding Banning removal of vegetation in areas susceptible to landslides Deforestation-free policies

(Continued)

TABLE 7.3 (Continued)

NON-STRUCTURAL	
Strategy domain	Cornerstones – Actions
Engagement in decision-making and practice	
Institutional frameworks and policy making	Establishment of legal and institutional frameworks directed to disaster risk reduction Reinforcement of legal and institutional frameworks directed to disaster risk reduction Informed or scientific-based policy making Mainstreaming landslide disaster risk reduction into development Implementing integrated landslide disaster risk management (ILDRiM) Formulating landslide policies at different geographical (i.e., catchment) and administrative scales (local, municipal, subnational, national) Enabling integrated-transdisciplinary participatory processes with all relevant stakeholders Enhancing integrated landslide community-based management and partnership Establishing permanent partnerships within and across sectors and levels of government Policy coordination Advice and advocacy frameworks to manage landslide disaster risk. Policies to avoid the construction of new housing, infrastructure, and facilities in areas prone to landslides Policies for reduction of population exposure to landslides Regulation of squatter settlement in landslide-prone areas Policy frameworks for good relocation of settlements located in highly susceptible landslide areas
Social cohesion	Local social structure strengthening Establishment and strengthening of local community committees Promoting community participation in landslide evacuation drills Enabling local landslide disaster risk action plans
Building codes and other codes of practice	Reinforcing and implementing of codes for the design, operation, and maintenance of housing, buildings, and infrastructure facilities in areas susceptible to landsliding Implementation of sanitary and other utility codes of practice Avoiding sewage effluents and limiting overflows Cleaning and maintenance of primary and secondary drainage channels and clogged cross-drainage works
Financing policies schemes	Implementation of insurance systems and coverages

Source: Alcántara-Ayala & Geertsema (2022)

Landslide Early Warning Systems

Predicting landslide triggers requires analyzing precursors, indicators, and rainfall thresholds (Sassa et al., 2009). Monitoring landslides is essential to comprehending their mechanisms and forecasting their occurrence, and it can be done through field observations, ground-based observation, and remote sensing methods. These methods provide critical information for landslide early warning systems (LEWSs) (Chae et al., 2017).

One of the earliest attempts to create a LEWS using modern science was in 1969 by Campbell (1975). He proposed integrating rain gauges to monitor hourly precipitation, a weather-mapping system to identify areas of high-intensity rainfall, and an administrative network to disseminate crucial information to residents. This approach would enable timely and proactive alerts to be issued, which could help mitigate the impact of severe weather events.

Similarly, from a technical perspective, LEWSs aim to provide timely and effective information to minimize damage and loss of life. Precursor forecasting, occurrence, event initiation, and impact are crucial (Picarelli et al., 2007). Early warning procedures based on precursor measurements are necessary for short timeframes. Rainfall is widely used for early warning of landslides (Sassa et al., 2009).

Inter-, intra-, and transdisciplinary approaches are essential for the integrated design and effective implementation of landslide early warning systems (LEWS) based on risk scenarios, not just hazard characterization (see Box 7.7). Most investigations have focused on characterizing landslide hazards and developing new technologies, models, and prototypes to improve the accuracy of understanding landslide mechanisms and spatial-temporal predictions. Additionally, the successful and sustained operation of LEWSs requires a transdisciplinary approach and an understanding of the social contexts in which they will be implemented (Alcántara-Ayala & Garnica-Peña, 2023a, 2023b).

Alcántara-Ayala and Garnica-Peña (2023c) argue that lack of resources is a challenge in implementing LEWSs. High equipment costs, vandalism, insecurity, and power outages hinder long-term usability. Collaboration with local authorities can be difficult due to bureaucracy and lack of interest while lacking human resources limits capacity building. Effective LEWSs require involvement from at-risk communities and society-wide action, including disaster risk communication and traditional oral knowledge (see Box 7.7).

Several researchers have addressed the architecture of early warning systems (EWSs) (Anderson, 1969; Grundfest et al., 1978; UN Global Survey of Early Warning Systems, 2006; Kelman & Glantz, 2014), questioning the assumptions behind implementing EWSs, which have primarily focused on technical aspects while disregarding disaster risk reduction policies and community involvement in knowledge processes. They focus on avoiding the consequences of vulnerability (including exposure) to nature and often emphasize the difficulties of predicting hazards and multi-hazard dynamics. Many existing early warning systems prioritize disaster impact reduction over disaster risk reduction, leaving communities vulnerable to future disasters.

LEWSs use instrumentation, monitoring, and communication channels to collect data and establish response protocols. By considering local perspectives, EWSs can be more effective in addressing real-world challenges (Box 7.8).

BOX 7.7 The Slope Safety System: Beyond Traditional LEWSs

For over four decades, the Geotechnical Control Office has been at the forefront of regulating geotechnical engineering practices in Hong Kong and ensuring that landslide risks are carefully monitored. Today, this office, now known as GEO, has put in place a comprehensive Slope Safety System that not only contains the risk of existing while reducing the consequences of any potential risks but also minimizes the risk of new developments (Pun et al., 2020). Hong Kong's approach to managing landslide risks emphasizes continuous improvement, with LEWSs, safety standards, and specialist works projects being key focus areas. The Geotechnical Engineering Office is vital in advancing emergency preparedness and crisis management during extreme weather events, and its dedication to public education and innovative technologies helps anticipate and mitigate potential hazards (Cheung, 2021).

Sources: Pun et al. (2020); Cheung (2021)

BOX 7.8 The Thai Landslide Network: A Commitment to Reduce Disaster Risk in the Tambons

Deforestation and settlement expansion in rural mountain areas of Thailand have increased the risk of landslides. Mapping of landslide-prone areas is based on geology, morphology, and vulnerability of people. Community-based observation networks train locals to detect early signs of landslides for timely evacuation. The Geohazards Centre coordinates between Thai agencies, villages, tambons (communities), and rescue services and collaborates with the Thai Meteorological Department for weather forecasting in landslide prevention. The Thai landslide network has successfully saved lives and educated and sensitized locals through volunteer training and hazard mitigation structures. Emphasis is placed on timely evacuations and educating loggers on the impact of illegal logging and deforestation on the community. Finding new sources of income is also necessary. By monitoring logging activities and planning land use more effectively, the risk of landslides can be reduced.

Source: Schmidt-Thomé et al. (2018)

Understanding the impact of landslides on the Earth's surface is essential for sustainable land use planning and hazard management, which along with the implementation of policy frameworks to reduce vulnerability, are indispensable for dealing with landslides (Alcántara-Ayala, 2021). Nonetheless, it is important to understand that landslides are more than hazards. They are fundamental in landscape evolution.

Korup (2012) investigates landslides in the Earth system, discussing that the impact of landslides on the Earth's surface results in the movement and redistribution of soil, rock, debris, and other materials. Likewise, he underlines that the magnitude of slope failures can range from minuscule to exceedingly large, spanning more than 26 orders of magnitude. As quantitative data on their distribution, scaling, and consequences become more readily available, it is possible to systematically synthesize their impact on the lithosphere, hydrosphere, and biosphere. The effects of landslides can be far-reaching and have a profound impact on the topography of bedrock. They can trigger shallow crustal response, limit volcanic edifice growth, alter water and sediment flow in river systems, and affect soil production rates. Moreover, they can rejuvenate and modify hillslope and riparian habitats.

Debris flows impact sediment and wood movement in mountainous basins of the Pacific Coastal Ecoregion (Montgomery & Buffington, 1998). The effect varies based on watershed attributes, such as geology, topography, soils, and hydrologic characteristics. Factors like channel gradient and tributary junction angles influence debris flow routing and recovery rates (Benda & Cundy, 1990). Landslides can cause local bed aggradation and channel type alterations, affecting riparian vegetation and the community composition (Naiman et al., 2000).

Similarly, low-order streams in mountainous regions pose a high propensity for erosion (Beschta & Platts, 1986), and human activities have significantly negatively impacted their delicate ecosystems. Logging practices on hillslopes that are prone to landslides exacerbate the issue by increasing debris flow, which often results in these small streams becoming bedrock channels (Montgomery et al., 1996). Wood recruitment can suffer greatly when streamside buffers are absent, or low-order basins are intensively harvested for timber. As a result, the streams' sediment storage capacity is reduced, and sediment is more likely to flow downstream, causing problems for other areas (May & Gresswell, 2003).

Foster et al. (2020) looked at the recovery of an ecosystem in southwestern Washington after a severe storm caused debris flows. The team monitored streams, biota recovery rates, physical habitat, and water temperature to identify factors affecting recovery. The research showed how the debris flows caused extensive damage and led to varying levels of resilience and recovery in aquatic species depending on physical habitat conditions and nearby source populations.

Conclusions

Understanding the risks associated with tsunamis and landslides requires interdisciplinary and transdisciplinary approaches. While there is still much uncertainty around the geodynamics of these hazards, it is important to focus on risk management through integrated and systematic efforts. This involves redefining warning systems to fit community realities and contexts, incorporating local risk perceptions and knowledge, adapting educational strategies for specific vulnerable groups, implementing building codes and construction standards, planning land use, protecting coastal areas, mountains, and slopes, and regularly updating risk assessments to improve urbanization and territorial planning processes.

References Cited

Alcántara-Ayala, I. (2021). Integrated landslide disaster risk management (ILDRiM): The challenge to avoid the construction of new disaster risk. *Environmental Hazards, 20*(3), 323–344. https://doi.org/10.1080/17477891.2020.1810609

Alcántara-Ayala, I., & Garnica-Peña, R. J. (2023a). Landslide warning systems in upper middle-income countries: Current insights and new perspectives. In I. Alcántara-Ayala, Z. Arbanas, D. Huntley, K. Konagai, M. Mikoš, K. Sassa, S. Sassa, H. Tang, & B. Tiwari (Eds.), *Progress in landslide research and technology* (Vol. 1, No. 2, pp. 159–168). Springer International Publishing/Springer Nature Switzerland AG. https://doi.org/10.1007/978-3-031-18471-0_13

Alcántara-Ayala, I., & Garnica-Peña, R. J. (2023b). Landslide warning systems in high-income countries: Past accomplishments and expected endeavours. In I. Alcántara-Ayala, Z. Arbanas, D. Huntley, K. Konagai, M. Mikoš, K. Sassa, S. Sassa, H. Tang, & B. Tiwari (Eds.), *Progress in landslide research and technology* (Vol. 2, No. 1). Springer International Publishing/Springer Nature Switzerland AG. https://doi.org/10.1007/978-3-031-39012-8_5

Alcántara-Ayala, I., & Garnica-Peña, R. J. (2023c). Landslide warning systems in low-and lower-middle-income countries: Future challenges and societal impact. In I. Alcántara-Ayala, Z. Arbanas, D. Huntley, K. Konagai, M. Mikoš, K. Sassa, S. Sassa, H. Tang, & B. Tiwari (Eds.), *Progress in landslide research and technology* (Vol. 1, No. 1, pp. 137–147). Springer International Publishing/Springer Nature Switzerland AG. https://doi.org/10.1007/978-3-031-16898-7_9

Alcántara-Ayala, I., & Geertsema, M. (2022). Landslides. In T. K. McGee & E. C. Penning-Rowsell (Eds.), *Routledge handbook of environmental hazards and society* (pp. 49–70). Routledge.

Anderson, W. A. (1969). Disaster warning and communication processes in two communities. *Journal of Communication, 19*(2), 92–104.

Assis Dias, M. C. D., Saito, S. M., Alvalá, R. C. D. S., Stenner, C., Pinho, G., Nobre, C. A., Fonseca, M. R. D. S., Santos, C., Amadeu, P., Silva, D., Lima, C. O., Ribeiro, J., Nascimento, F., & Corrêa, C. D. O. (2018). Estimation of exposed population to landslides and floods risk areas in Brazil, on an intra-urban scale. *International Journal of Disaster Risk Reduction, 31*, 449–459. https://doi.org/10.1016/j.ijdrr.2018.06.002

Benda, L. E., & Cundy, T. W. (1990). Predicting deposition of debris flows in mountain channels. *Canadian Geotechnical Journal, 27*(4), 409–417. https://doi.org/10.1139/t90-057

Beschta, R. L., & Platts, W. S. (1986). Morphological features of small streams: Significance and function 1. *JAWRA Journal of the American Water Resources Association, 22*(3), 369–379. https://doi.org/10.1111/j.1752-1688.1986.tb01891.x

Boschetti, L., & Ioualalen, M. (2021). Integrated tsunami intensity scale based on maxima of tsunami amplitude and induced current. *Natural Hazards, 105*, 815–839. https://doi.org/10.1007/s11069-020-04338-5

Brunsden, D. (1979). Mass movements. In C. E. Embleton & J. B. Thornes (Eds.), *Process in geomorphology* (pp. 130–186). Edward Arnold.

Bryant, E. A., & Nott, J. (2001). Geological indicators of large tsunami in Australia. *Natural Hazards, 24*, 231–249. https://doi.org/10.1023/A:1012034021063

Campbell, R. H. (1975). *Soil slips, debris flows, and rainstorms in the Santa Monica Mountains and vicinity, southern California. U.S. geological survey professional paper 851*. United States Government Printing Office. https://pubs.usgs.gov/pp/0851/report.pdf

Chae, B. G., Park, H. J., Catani, F., Simoni, A., & Berti, M. (2017). Landslide prediction, monitoring and early warning: A concise review of state-of-the-art. *Geosciences Journal, 21*, 1033–1070. https://doi.org/10.1007/s12303-017-0034-4

Cheung, R. W. (2021). Landslide risk management in Hong Kong. *Landslides, 18*(10), 3457–3473. https://doi.org/10.1007/s10346-020-01587-0

Clague, J. J., & Stead, D. (Eds.). (2012). *Landslides: Types, mechanisms and modeling*. Cambridge University Press.

Cruden, D. M., & Varnes, D. J. (1996). Landslide types and processes. In A. K. Turner & R. L. Schuster (Eds.), *Landslides: Investigation and mitigation* (pp. 36–75). Transportation Research Board Special Report, National Academy of Sciences, Special Report 247. https://onlinepubs.trb.org/Onlinepubs/sr/sr247/sr247.pdf

Dourado, F., Arraes, T. C., & Silva, M. F. (2012). The "megadesastre" in the mountain region of Rio de Janeiro State: Causes, mechanisms of mass movements and spatial allocation of investments for reconstruction post disaster. *Anuário do Instituto de Geociências, 35*(2), 43–54.

Ferris, E., & Solis, M. (2013, March 11). *Earthquake, tsunami, meltdown – the triple disaster's impact on Japan, impact on the world.* The Brookings Institution. https://www.brookings.edu/articles/earthquake-tsunami-meltdown-the-triple-disasters-impact-on-japan-impact-on-the-world/

Foster, A. D., Claeson, S. M., Bisson, P. A., & Heimburg, J. (2020). Aquatic and riparian ecosystem recovery from debris flows in two western Washington streams, USA. *Ecology and Evolution, 10*(6), 2749–2777. https://doi.org/10.1002/ece3.5919

Fundação SOS Mata Atlântica. (2018). *Atlas dos Remanescentes Florestais da Mata Atlântica período 2016–2017.* Fundação SOS Mata Atlântica, Instituto Nacional de Pesquisas Espaciais, São Paulo. https://cms.sosma.org.br/wp-content/uploads/2021/05/SOSMA_Atlas-da-Mata-Atlantica_2019-2020.pdf

Gaillard, JC., Clavé, E., Vibert, O., Azhari, Dedi, Denain, J. C., Efendi, Y., Grancher, D., Liamzon, C. C., Rosnita Sari, D., & Setiawan, R. (2008). Ethnic groups' response to the 26 December 2004 earthquake and tsunami in Aceh, Indonesia. *Natural Hazards, 47*, 17–38. https://doi.org/10.1007/s11069-007-9193-3

Geertsema, M., & Alcántara-Ayala, I. (2022). Mountain landslides: An overview of common types and future impacts. In F. O. Sarmiento (Ed.), *Montology palimpsest: A primer of mountain geographies* (Vol. 1, pp. 187–209). Springer Nature.

Grundfest, E. C., White, G. F., & Downing, T. C. (1978). Big Thompson flood exposes need for better flood reaction system to save lives. *Civil Engineering – ASCE, 48*(2), 72–73.

Guerra, A. J. T., Lopes, P. B. M., & dos Santos Filho, R. D. (2007). Características Geográficas e Geomorfológicas da APA, Petrópolis, RJ. *Revista Brasileira de Geomorfologia, 8*(1).

Gupta, H. K., & Gahalaut, V. K. (2013). *Three great tsunamis: Lisbon (1755), Sumatra-Andaman (2004) and Japan (2011): Springer briefs in earth sciences.* Springer.

Harada, N., Shigemura, J., Tanichi, M., Kawaida, K., Takahashi, S., & Yasukata, F. (2015). Mental health and psychological impacts from the 2011 Great East Japan earthquake disaster: A systematic literature review. *Disaster and Military Medicine, 1*(1), 1–12. https://doi.org/10.1186/s40696-015-0008-x

Highland, L., & Bobrowsky, P. T. (2008). *The landslide handbook: A guide to understanding landslides.* US Geological Survey No. 1325. pubs.usgs.gov/circ/1325/pdf/C1325_508.pdf

Hills, J. G., & Mader, C. L. (1997). Tsunami produced by the impacts of small asteroids. *Annals of the New York Academy of Sciences, 822*(1), 381–394. https://doi.org/10.1111/j.1749-6632.1997.tb48352.x

Hungr, O., Leroueil, S., & Picarelli, L. (2014). The Varnes classification of landslide types, an update. *Landslides, 11*, 167–194. https://doi.org/10.1007/s10346-013-0436-y

Igarashi, Y., Kong, L., Yamamoto, M., & McCreery, C. S. (2011). Anatomy of historical tsunamis: Lessons learned for tsunami warning. *Pure and Applied Geophysics, 168*(11), 2043–2063. https://doi.org/10.1007/s00024-011-0287-1

Jonathan, M. P., Godwyn-Paulson, P., Rodríguez-Espinosa, P. F., & Roy, P. D. (2023). Tsunami waves across the Pacific ring nations from January 15, 2022, underwater volcanic explosion at "the Hunga-Tonga-Hunga-Ha'apai" isles. *Natural Hazards, 115*(2), 1861–1871. https://doi.org/10.1007/s11069-022-05583-6

Kelman, I., & Glantz, M. H. (2014). Early warning systems defined. In Z. Zommers & A. Singh (Eds.), *Reducing disaster: Early warning systems for climate change* (pp. 89–108). Springer.

Klug, H. (1986). *Flutwellen und Risiken der Küste*. Franz Steiner Verlag.

Korup, O. (2012). Landslides in the earth system. In J. J. Clague & D. Stead (Eds.), *Landslides: Types, mechanisms and modeling* (pp. 10–23). Cambridge University Press.

Kurkin, A., Belyakov, V., Makarov, V., Zeziulin, D., & Pelinovsky, E. (2016). Methods of tsunami detection and of post-tsunami surveys. *Science of Tsunami Hazards, 35*(2), 68–83.

Leroueil, S., Locat, J., Vaunat, J., Picarelli, L., Lee, H., & Faure, R. M. (1996). Geotechnical characterization of slope movements. In K. Senneset (Ed.), *Landslides* (pp. 53–74). Balkema.

Levin, B. W., & Nosov, M. A. (2016). *Physics of tsunamis*. Springer.

Marchezini, V., & Wisner, B. (2017). Challenges for vulnerability reduction in Brazil: Insights from the PAR framework. In V. Marchezini, B. Wisner, L. R. Londe, & S. M. Saito (Eds.), *Reduction of vulnerability to disasters: From knowledge to action* (pp. 57–96). Editora Rima.

May, C. L., & Gresswell, R. E. (2003). Processes and rates of sediment and wood accumulation in headwater streams of the Oregon Coast Range, USA. *Earth Surface Processes and Landforms, 28*(4), 409–424. https://doi.org/10.1002/esp.450

McAdoo, B. G., Dengler, L., Prasetya, G., & Titov, V. (2006). Smong: How an oral history saved thousands on Indonesia's Simeulue Island during the December 2004 and March 2005 tsunamis. *Earthquake Spectra, 22*(Suppl. 3), 661–669. https://doi.org/10.1193/1.2204966

Mikulecký, P., Punčochářová, A., Babič, F., Bureš, V., Čech, P., Husáková, M., Mls, K., Nacházel, T., Ponce, D., Štekerová, K., Triantafyllou, I., Tučník, P., Sunanda, V., & Zanker, M. (2023). Dealing with risks associated with tsunamis using indigenous knowledge approaches. *International Journal of Disaster Risk Reduction, 86*, 103534. https://doi.org/10.1016/j.ijdrr.2023.103534

Montgomery, D. R., Abbe, T. B., Buffington, J. M., Peterson, N. P., Schmidt, K. M., & Stock, J. D. (1996). Distribution of bedrock and alluvial channels in forested mountain drainage basins. *Nature, 381*(6583), 587–589. https://doi.org/10.1038/381587a0

Montgomery, D. R., & Buffington, J. M. (1998). Channel processes, classification, and response. In R. Naiman & R. Bilby R (Eds.), *River ecology and management: Lessons from the Pacific coastal ecoregion* (pp. 13–42). Springer-Verlag.

Naiman, R. J., Bilby, R. E., & Bisson, P. A. (2000). Riparian ecology and management in the Pacific coastal rain forest. *BioScience, 50*(11), 996–1011. https://doi.org/10.1641/0006-3568(2000)050[0996:REAMIT]2.0.CO;2

Owen, J. (2005, January 18). Tsunami family saved by schoolgirl's geography lesson. *National Geographic News*. https://ceas.uchicago.edu/sites/ceas.uchicago.edu/files/uploads/sti2013_arehart_nationalgeographic_tillysmith_tsunami.pdf

Pararas-Carayannis, G. (2022). Hunga-Tonga-Hunga-Ha'pai volcanic eruption/explosion and tsunamis of 14–15 January 2022. *Science of Tsunami Hazards, 41*(3), 286–310.

Pattiaratchi, C. B., & Wijeratne, E. M. S. (2015). Are meteotsunamis an underrated hazard? *Philosophical Transactions of the Royal Society A: Mathematical, Physical and Engineering Sciences, 373*(2053), 20140377. https://doi.org/10.1098/rsta.2014.0377

Picarelli, L., Versace, P., Olivares, L., & Damiano, E. (2007). *Prediction of rainfall induced landslides in unsaturated granular soils for early warning setting up*. Proceedings of 2007 International Forum on Landslide Disaster Management, Hong Kong.

Pun, W. K., Chung, P. W. K., Wong, T. K. C., Lam, H. W. K., & Wong, L. A. (2020). Landslide risk management in Hong Kong-experience in the past and planning for the future. *Landslides, 17*, 243–247. https://doi.org/10.1007/s10346-019-01291-8

Röbke, B. R., & Vött, A. (2017). The tsunami phenomenon. *Progress in Oceanography, 159*, 296–322. https://doi.org/10.1016/j.pocean.2017.09.003

Rounce, D. R., Hock, R., Maussion, F., Hugonnet, R., Kochtitzky, W., Huss, M., Berthier, E., Brinkerhoff, D., Compagno, L., Copland, L., Farinotti, D., Menounos, B., & McNabb, R. W. (2023). Global glacier change in the 21st century: Every increase in temperature matters. *Science, 379*(6627), 78–83. https://doi.org/abo1324

Sassa, K., Picarelli, L., & Yueping, Y. (2009). Monitoring, prediction and early warning. Landslides–disaster risk reduction. In K. Sassa & P. Canuti (Eds.), *Landslides – disaster risk reduction* (pp. 351–375). Springer. https://doi.org/10.1007/978-3-540-69970-5_20

Schmidt-Thomé, P., Tatong, T., Kunthasap, P., & Wathanaprida, S. (2018). Community based landslide risk mitigation in Thailand. *Episodes Journal of International Geoscience, 41*(4), 225–233. https://doi.org/10.18814/epiiugs/2018/018017

Self, S., & Rampino, M. R. (1981). The 1883 eruption of Krakatau. *Nature, 294*(5843), 699–704. https://doi.org/10.1038/294699a0

Skempton, A. W., & Hutchinson, J. (1969). Stability of natural slopes and embankment foundations. In *Proceedings, 7th international conference of soil mechanics and foundation engineering* (pp. 291–340). State of the Art Volume.

Sutton, S. A., Paton, D., Buergelt, P., Sagala, S., & Meilianda, E. (2020). Sustaining a transformative disaster risk reduction strategy: Grandmothers' telling and singing tsunami stories for over 100 years saving lives on Simeulue Island. *International Journal of Environmental Research and Public Health, 17*(21), 7764. https://doi.org/10.3390/ijerph17217764

Tappin, D. R. (2018). The importance of geologists and geology in tsunami science and tsunami hazard. *Geological Society, London, Special Publications, 456*(1), 5–38. https://doi.org/10.1144/SP456.11

Tappin, D. R., Grilli, S. T., Harris, J. C., Geller, R. J., Masterlark, T., Kirby, J. T., Shi, F., Ma, G., Thingbaijam, K., & Mai, P. M. (2014). Did a submarine landslide contribute to the 2011 Tohoku tsunami? *Marine Geology, 357*, 344–361. https://doi.org/10.1016/j.margeo.2014.09.043

Terzaghi, K. (1950). *Mechanism of landslides* (Berkey Vol.). Geological Society of America.

Tkalich, P. (2011). Evolution of tsunami science. *Asia Pacific Mathematics Newsletter, 1*(3), 11–16.

UN. (2006). *Global survey of early warning systems: An assessment of capacities, gaps and opportunities towards building a comprehensive global early warning system for all natural hazards.* United Nations. https://www.unisdr.org/2006/ppew/info-resources/ewc3/Global-Survey-of-Early-Warning-Systems.pdf

Varnes, D. J. (1978). Slope movement types and processes. *Special Report, 176*, 11–33. https://onlinepubs.trb.org/onlinepubs/sr/sr176/176-002.pdf

Vučetić, T., Vilibić, I., Tinti, S., & Maramai, A. (2009). The great adriatic flood of 21 June 1978 revisited: An overview of the reports. *Physics and Chemistry of the Earth, Parts A/B/C, 34*(17–18), 894–903. https://doi.org/10.1016/j.pce.2009.08.005

Wünnemann, K., & Weiss, R. (2015). The meteorite impact-induced tsunami hazard. *Philosophical Transactions of the Royal Society A: Mathematical, Physical and Engineering Sciences, 373*(2053), 20140381. https://doi.org/10.1098/rsta.2014.0381

Zaruba, Q., & Mencl, V. (2014). *Landslides and their control.* Elsevier.

Zeng, H., Tang, C. S., Cheng, Q., Zhu, C., Yin, L. Y., & Shi, B. (2020). Drought-induced soil desiccation cracking behavior with consideration of basal friction and layer thickness. *Water Resources Research*, 56(7), e2019WR026948.

Chapter Eight
Beyond the Earth

Our planetary home sits in a vast universe. Our home galaxy, the Milky Way, is situated within a massive supercluster of galaxies as a tiny speck. Our solar system rotates around the galaxy's center in an outer area of one arm of the galaxy. The Earth, a rare and precious habitat for life as described in Chapter 4, is merely a medium-sized object within this solar system. Hazards affect this inconsequential object from beyond it.

Our solar system's star, the Sun, is a main source of energy for all biotic and abiotic processes taking place on, surrounding, and within the Earth. The Earth's movement around the Sun and the position of its axis determine seasons, while solar radiation drives the weather and climate (Chapter 5). Our proximity to the Moon contributes to the tides, creating zones between land and water of life and of hazard. The death explosions of stars over eons created and distributed all the heavy elements that make up Earthly life and minerals. The fiery origins of the solar system and early collisions as planets established their orbits provided Earth with a rotating molten iron core that generated a magnetic field that protects life from harmful radiation from space. Other future planetary collisions or other hazards with origins from beyond the Earth could significantly interfere with and harm Earthly life – even lead to mass extinctions. This chapter examines some of these hazards emerging from beyond our cocoon on Earth.

The Earth as a Least Hazardous Planet?

Before examining hazards from beyond the Earth, consider the relatively rare conjuncture of conditions making the Earth hospitable to life and its evolution. Calculations regarding the expectation of life on a planet orbiting a star can be traced back to at least Huang

DOI: 10.4324/9781003292814-8

(1959a) further developed in Huang (1959b), who coined and examined the "habitable zone" for life. These papers recognized the difficulties in defining "life" and in assuming that life must be similar to that found on Earth yet represent a solid and important beginning. Huang (1959a, p. 397) even made the optimistic prediction that "In a few years [humanity] will learn with certainty about the existence or non-existence of life on the other planets in the solar system", a question which, at the time of writing this book, remained open (Bains et al., 2023).

With the "habitable zone" baseline for understanding what areas of the universe Earth-like life might exist, the search for it intensified. Project Ozma (Drake, 1961) outlined equipment to look for radio signals that might indicate life. The Drake Equation (Drake, 1965, p. 324) followed, estimating the number of "communicative civilisations" – that is, life forms using technology that can be detected over interstellar distances, with radio waves being a prime example. This equation finds this number (N) by multiplying the factors:

- Rate of star formation (R_*);
- The fraction of these stars with planets (f_p);
- The mean number of planets per star with life-permitting environments (n_e);
- The fraction of these planets on which life occurs (f_l);
- The fraction of these planets on which intelligent life occurs (f_i);
- The fraction of these planets on which the intelligent life produces a communicative civilization (f_c); and
- The communicative civilization's mean lifetime (L).

In equation form, $N = R_* f_p n_e f_l f_i f_c L$, where N is the number of "communicative civilisations".

Around the late 1990s or perhaps slightly later, although the coining is obscure, the "habitable zone" became the "Goldilocks Zone" or "The Goldilocks Effect". The fairy tale "The Three Bears" involves a girl named Goldilocks finding the possessions of three bears. One set is always too much, one set is always too little, and one set is just perfect for her. From the analogy of these bears, planets too far from a star would be too cold for Earth-like life, planets too close to a star would be too hot for Earth-like life, and planets in between would be just the right temperature, or in the Goldilocks Zone. Similarly, analogies might apply to aspects of the size and shape of a planet required for intelligent life to develop as well as other variables, such as nutrients available and the rate of environmental change compared to the life's lifetimes and evolution rate. All these analyses remain focused on Earth-like life.

One other postulated point relates the Drake Equation's factor *L*, the lifetime of a civilization that is detectable at interstellar distances. Aside from natural forces causing extinctions, how many intelligent civilizations end up in a self-destructive mode? For human beings, nuclear war and runaway climate change triggered by human actions, making the planet too hot or too cold, are suggestions of mechanisms to knock us out of Drake's "communicative" state, even without human extinction. Speculations suggest that intelligent civilizations must go through one or more self-destructions before achieving success in moving off-planet without ruining the place of their origins. For now, humanity is working with an empirical sample size of one, for which the trajectory toward or away from self-destruction has yet to be determined.

In the meantime, numerous planets with life-permitting environments have been discovered, affecting the estimate of the Drake Equation's variable f_p. Exoplanets (also called exosolar planets) are planets outside our solar system, first confirmed by Wolszczan and Frail (1992) despite earlier candidates. At the time of writing, over 5,000 exoplanets were accepted with nearly twice as many possible candidates. Exoplanet discovery was followed by the first planet found orbiting a Sun-type star (Mayor & Queloz, 1995) and then, for factor n_e in the Drake Equation, the first planet found in the Earth-like habitable zone (Santos et al., 2001). What about planets with life and with life able to deal with all the hazards from their environment? We await that discovery. Planet Earth, at least so far, has not been hazardous enough to preclude life or to completely destroy life that has developed.

Hazards From Outer Space

We gaze up at the night sky and witness a streak of light blazing across the dark dome. Perhaps we make a wish, perhaps we reminisce about the dinosaurs who might have seen a similar phenomenon as their last view, or perhaps we simply enjoy the moment an object from outer space enters the Earth's atmosphere. Most of these instances involve small near-Earth objects (NEOs) heading toward us and burning up when entering the atmosphere. Some objects are so big or are moving so fast that they do not burn up completely before reaching the ground. The impact can be cataclysmic. Such event might have caused mass extinctions, notably the one to which non-bird dinosaurs succumbed according to one school of thought – with flood basalts, large lava releases (Chapter 6), being another hypothesis, as well as possible combinations (Schoene et al., 2019; Sprain et al., 2019).

Today around the world, we map out the edges of craters with diameters of dozens of kilometers. Meteor Crater Natural Landmark in Arizona, USA, is an exciting tourist attraction with lookout points, guided tours around the rim, movie theaters, a space museum, and a science discovery center. It was also a training site for humanity's first steps on the Moon. Far from tourist trails, many other craters remain underneath the ocean or have vanished due to tectonic plate movement (Chapter 6).

NEOs come in three main forms. *Asteroids* are chunks of rocks in a wide variety of highly irregular shapes, ranging in size from meters to hundreds of kilometers across. *Comets* are similar, although their core is usually icy and enveloped by gas and dust which streams away from the Sun, like a tail up to millions of kilometers long. *Meteoroids* are smaller than asteroids or comets, presumed to be remnants or tiny chunks of them, often no bigger than pebbles. Most NEOs orbit around the Sun, with over 1 million known asteroids and nearly 3,000 known comets. The vast majority of meteoroids are too small to detect unless they are very close to the Earth. Various bands and clouds of these objects exist in and around our solar system, with the likelihood of their numbers being in the billions. Space objects do come from outside the solar system and might be captured by the Sun's or a planet's gravity – or they might swing by once and then depart.

The NEO's size and speed are significant factors determining the object's likely consequences if it enters the Earth's atmosphere and if it impacts the planet's surface. Thousands of NEOs with the capability of causing global-scale damage have been discovered. For the moment, it does not appear as if any will come close to the Earth for decades, if not longer. NEO trackers update probability impacts as more data are available while issuing alerts when an object is known to be on a collision course with the Earth – most frequently so that people can enjoy watching it burn up in the atmosphere.

These efforts show how, while the consequences of a large NEO strike would be devastating, society is capable of identifying and responding to the potential threats (Brobrowsky & Rickman, 2007). In particular, the worse the threat, the easier it typically is to know about in advance, as long as we are looking. Generally, the more massive a space object and the faster it moves, the larger its impact on the planet, but the quicker we could discover it. Most of the NEOs that we know about have a reasonable similarity in their compositions so that a larger mass tends to mean a larger volume which is easier to spot. It is the same with speed. The faster an object, the more obvious its movements are against the backdrop of comparatively fixed stars and so the easier it is to detect and calculate its trajectory.

A large NEO does not even need to reach the Earth's surface in order to wipe out cities. A large and/or swift space object on a path taking it through the atmosphere without hitting the surface might skip through the upper layers burning off material. Or it could explode. Either way, a massive shockwave of pressure and heat could blast the Earth's surface. If this shockwave hits inhabited places, then zones of complete destruction and death could be surrounded by areas of mass casualties and damage. A large NEO pummeling the Earth's surface is preceded by a wave of pressure and heat followed by the blast of impact annihilating anything underneath, sending ripples across the Earth's surface that exceed the largest recorded earthquakes, and generating an expanding dust cloud that might circle the planet, block some of the Sun's heat, and cool the climate for years. Food supplies and supply chains would be in disarray, inducing mass starvation. If such a NEO hits water, more likely than hitting land since oceans make up 71% of the Earth's surface, a huge tsunami could form (Chapter 6). A mid-ocean impact could inundate coastlines with waves over a hundred meters high reaching dozens of kilometers inland. Could we evacuate millions of coastal dwellers far enough away within hours? A near-shore impact might not give coastal inhabitants time to register and comprehend the shockwave or water tower before being obliterated.

Such destruction may or may not have occurred several thousand years ago, with the date disputed by centuries, when a space object smashed into Estonia's island of Saaremaa, creating the Kaali craters visible today. The date of the first human habitation on the island is debated as much as the date of the impact, meaning that human settlements might have been annihilated, but it is hard to confirm exactly what happened.

More definitive are records of the blast above Siberia's Tunguska region on 30 June 1908. What exactly happened has never been determined, although it is presumed to have been from outer space since the human capability for explosives at that time was not sufficient to produce the observations. Considering that neither an impact crater nor pieces of an object have been confirmed, the most likely explanation is a meteor that pulverized before it impacted the ground. Trees bent and snapped across a wide radius with damage increasing concentrically toward the location of the presumed explosion. A man dozens of kilometers away was thrown from his chair by the blast. Local folklore claims that people in the area were never seen again and so must have been incinerated.

Just over a century later, the Chelyabinsk region nearly suffered a similar fate (Popova et al., 2013). With ubiquitous dashcams, phones, and security cameras capturing the images, a meteor brightened the dawn of 15 February 2013. Smashed window glass injured over 1,200 people. Rock fragments were scooped up all around, without confirmed reports of people being hit. This incident is the first definitive mass casualty event from a NEO.

Asteroids measuring at least 50 m in diameter are estimated to strike the Earth every millennium or so (Brown et al., 2002). In comparison, objects similar in size to the Chelyabinsk asteroid (around 20 m in diameter) hit the Earth every 50–100 years. Meanwhile, about 100 tons of small debris impact the Earth daily, although the effect is barely noticeable since it burns up before reaching the ground.

With knowledge of these previous instances and facing the prospect for future ones, what should we do? Reducing planet-wide vulnerability is not feasible nor would it seem to be helpful to presume that we could take the hit from a large NEO and clean up afterwards. Instead, actions could modify the hazard. One dramatic approach often depicted in fiction is destroying the NEO with nuclear weapons. This approach is feasible and could be considered an option of last resort, while being aware of the dangers. If anything goes wrong when deploying the nuclear bombs in space, parts of Earth could be contaminated. If the space object fragments, then large chunks could still pose dangers, possibly with the addition of some radioactivity.

Many other less destructive and more reliable actions have been proposed, focusing on deflecting the space object rather than destroying it. The sooner an object is deflected from its Earth-bound trajectory, the less of a nudge is required since a small deviation far away adds up over long distances. Suggestions include the following:

- Using spacecraft to push or pull the object, like a tugboat.
- Attaching an engine to the object to change its trajectory.
- Landing a large tank of water or other substance on the object's surface and then piercing the tank so that fluid venting into the vacuum of outer space would provide force to move the NEO.
- Painting parts of the object black to absorb sunlight and/or white to reflect sunlight, thus permitting differential heating to be the nudge that shifts the NEO's course.

The feasibility of these strategies depends on acting now (see the Box 8.1). After all, if humanity desires a serious program of surveillance, monitoring, and response only once a serious threat has been identified, then it will be too late. Preparation means being ready to go as soon as a threat is confirmed, noting the long lead time required to cover the distances across outer space. Since the more dangerous objects – the more massive and faster moving ones – are easier to identify, continuous, full-sky monitoring coupled with a variety of ready-to-go response systems could save humanity from a NEO-based calamity.

BOX 8.1 Near-Earth Object Monitoring, Warning, and Response

Several programs around the world aim to observe and map NEOs, while amateur astronomers scan the night sky and report their findings ad hoc. Funding is always difficult since it is long-term work forever adding to knowledge but not with immediate, demonstrable outcomes of having helped humanity – perhaps until it is too late.

In Wales, SPACEGUARD is the National Near Earth Objects Information Centre (NNEOIC). As an observatory and information center tracking NEOs, it communicates issues to the media and public while offering tours to visitors. Most of its funding comes from tour fees and donations.

Headquartered in Paris, the European Space Agency (ESA) currently has 22 member states working together on all these topics. ESA's Planetary Defence Office looks for NEOs, makes the data available publicly, and advises on actions when an object is deemed to be dangerous. As part of the ESA's wider activities, space missions are involved in testing the deflection of an asteroid through the impact of a spacecraft. Plans for space stations consider their use as a launching point for save-the-Earth-from-a-NEO.

In the US, the National Aeronautics and Space Administration (NASA) is planning a 2027 launch for its infrared telescope, the Near-Earth Object Surveyor space telescope (NEO Surveyor). Within the first five years, the scientists hope to identify at least two-thirds of the NEOs that could cause regional damage. In the meantime, as part of its NEO Observations Program, NASA supports programs across the US contributing to "planetary defense" through monitoring NEOs and being ready for when a threat is recognized.

The Japan Aerospace Exploration Agency (JAXA) tracks and monitors asteroids approaching the Earth through different observation centers, such as the Bisei Space Guard Center (BSGC) and Kamisaibara Space Guard Center (KSGC). They also follow orbital debris from broken satellites.

All of these responses have limits. If an object is big enough, we might have no prospect of stopping it. Another planet or even a Moon-sized object ambling into our solar system's neighborhood might be neither nudge-able nor destroyable. It might not even have to collide with our planet or, depending on its mass, come especially near. A massive-enough body would have sufficient gravity to alter the Moon's orbit around the Earth, rejigging our tides with significant knock-on effects for our planet's air-land-water interactions. The Earth, too, could be knocked out of orbit, ending up closer to the Sun so that we fry or farther from the Sun so that we freeze. The irony of experiencing an object of this size is that we should have plenty of warning time to know that we can do nothing about our impending doom.

Particles (and Rays) From Outer Space

Solar Storms

Charged particles are always heading toward us from the Sun in a steady stream called the solar wind. They can interfere with our electricity and communication systems, causing havoc for a society ill-prepared for outages of power, internet, and mobile phones. Fortunately, most of these particles are stopped by the magnetic field surrounding the Earth. The solar wind is channeled toward the planet's two magnetic poles, where they hit the atmosphere, causing the gorgeous dance of the southern and northern lights, the aurorae.

When the Sun ramps up its activity, large bursts of particles can turn the solar wind into a solar storm, also referred to as a geomagnetic storm. One form is a coronal mass ejection (or CME) in which the Sun explosively ejects large magnetic clouds, usually moving faster than the solar wind, with numerous other related phenomena sending charged particles toward the Earth (Tsurutani et al., 2022). Collectively known as space weather, these storms interfere with satellites, disrupting their electricity and impeding communication with them. Many satellites have some shielding to protect them, while predictions of and warnings for large solar storms can permit satellites to be shut down temporarily. This response supports the satellites being operational after the storm has passed yet means that the satellites cannot be used during the shutdown.

The power of geomagnetic storms alters the Earth's magnetic field, bringing aurorae to much lower latitudes. Highly energetic particles can punch right through, reaching ground level. This is when other day-to-day operations can be disrupted, with numerous examples emerging as soon as we started using electricity. Use of human-generated electricity began during the first half of the 18th century, leading to the invention and application of the telegraph 80 years later. The telegraph transmits electric signals along a wire permitting rapid communication between any two linked points. The first telegraph message was sent on 24 May 1844. Just 15 years later, on 1–2 September 1859, one of the largest solar storms yet observed directly hit the Earth. Named after Richard Carrington, who used his amateur astronomy skills to link observations on Earth with what he saw happening on the Sun, this Carrington Event stopped the functioning of several telegraph lines. Those relying on the telegraph lines experienced the first recorded space weather disaster. Others might not have been recorded. After all, we do not know how many ships at sea may have foundered off course, never heard from again, because their compasses were affected by solar storms.

The decades following the Carrington Event witnessed growing dependence on electricity to light our homes and settlements alongside communications through radio and telephone, eventually leading to the satellite communications era. This time period saw more solar storm outages. Southern Ontario and the northeastern USA were badly affected on 24 March 1940. Québec went through a major blackout in March 1989 due to several storms – with the lack of lights providing the opportunity to enjoy the magical aurorae! A series of space weather events in October and November 2003 affected the world, wrecking one satellite and interfering with many others as surface electricity and communications systems were disrupted.

At this point, the dangers to human health were also known. Responses to solar weather warnings rerouted commercial flights away from high latitudes, where the highest concentration of particles would enter the atmosphere. Astronauts also need to be cautious about their exposure. On Earth, Alabdulgader et al. (2018) report evidence that sharp or sudden variations in geomagnetic and solar activity as well as geomagnetic storms can stress the human body and be observed through changes in melatonin/serotonin balance, blood pressure, breathing, and heart rate as well as responses from the brain, the immune system, and the reproductive system.

So far, 23 July 2012 remains the closest that we have come to a global geomagnetic storm disaster. The storm was around the magnitude of the 1859 Carrington Event and was aimed right through the Earth's orbit. Fortuitously, our planet had passed the area of highest impact just one week earlier. A direct hit would have likely ruined most of our communications and electricity systems. Years would have been required to restore fried transformers and to repair or replace a good proportion of the satellites, which are now essential to societal functioning. The luck of seven days was all that averted a calamity costing trillions of US dollars.

As we learn more, we ramp up our preparations and warnings. Retrofitting and repairing systems is expensive and difficult, so preparedness and prevention should be incorporated into the design of electricity, information, and communication systems. Stopping particles hitting key components through electric and magnetic shielding is one approach. It rarely obstructs all space weather, so having backups and the capacity to shut down and restart swiftly, and at short notice, minimizes disruption. For instance, hospitals and emergency operations centers are generally connected to the main electricity grid which could be knocked out by a solar storm. Consequently, they have local generators or other backup power supplies and should ensure that they are well-shielded for continued functioning of life-saving equipment and procedures until the main grid is restored.

These design features apply equally to satellites and electricity supply systems. Transmission lines and internet communications are sometimes placed underground, which provides some shielding. There are disadvantages as well as advantages of underground placement. Underground cables are less of an eyesore than those on the surface, yet they require digging for installation and maintenance. They are less prone to ice storm and vehicle damage but are more prone to damage in heat waves and from irresponsible digging.

For solar storms, warnings are feasible on different time scales. The Sun goes through an 11-to-12-year cycle of maximum and minimum particle ejection. Periods of expected large storms can be known well in advance. Even at a low point in that cycle, a huge CME or solar flare could emerge at any time with only days of advance notice that something might happen, followed by only hours beforehand of the near-certainty of a storm. A definitive warning of a geomagnetic storm may only be provided minutes before it strikes. Rather than waiting until major space weather is already on its way, long-term readiness and willingness to act to protect infrastructure on the basis of forecasted impacts is required.

A single solar storm typically lasts hours, although periods of high activity can occur for days and waves may continue over weeks. During this period, airplanes dropping out of the sky and phones exploding are not expected. Global Positioning System (GPS) and communications could go out, so aircraft pilots, ship captains, and everyone else would need to rely on old-fashioned navigation to reach safety. Power surges could cause damage to equipment plugged into mains power. Lack of power would inhibit battery-operated devices from recharging.

Ultraviolet Radiation

The sun emits radiation of various energy levels ranging from long wave and lower energy infrared through to visible light and then to shorter wavelengths and higher energy ultraviolet (UV). UV radiation can cause skin cancers, cataracts, and immune system damage (WHO, 2023; EPA, 2022).

In one layer of the atmosphere, the stratosphere, ozone blocks some of the Sun's UV radiation from reaching the planet's surface, in effect preventing UV radiation's adverse impacts on life. A chemical, chlorofluorocarbons, was used in refrigerators and air conditioners as coolants, for industrial cleaning, and in aerosol spray cans. When chlorofluorocarbons are released into the air, they rise to the stratosphere, react with ozone, and deplete it. The thinner stratospheric ozone layer permits more UV rays to penetrate to the surface, affecting life. This effect was theorized by Molina and Rowland (1974) but not fully accepted until Farman et al. (1985) announced that they had identified significant ozone layer depletion over Antarctica. Given the realization that the problems could soon extend around the world, earlier action plans and conventions galvanized the signing of an international treaty that would eliminate chlorofluorocarbons (UN, 1987).

The ozone layer depletion, termed the "ozone hole", still appears over Antarctica every spring but closes over summer, so the situation is improving. Current expectations are that stratospheric ozone will return to its pre-chlorofluorocarbon baseline levels around 2050, depending on factors such as the illegal manufacture and release of chlorofluorocarbons and the release rates into the atmosphere of other gasses.

Cosmic Rays

Other particles, termed "cosmic rays", bombard us from the Sun and the rest of the universe. They are highly energetic nuclei of atoms, 89–90% of which are single protons, so nuclei of hydrogen. Helium nuclei with two protons comprise 9–10%, and the remainder can be the nucleus of any naturally occurring element or other particles, including electrons.

Cosmic rays can damage the cells of living beings, leading to cancer, which is a particular concern for astronauts spending a lot of time in space, including for interplanetary trips (Slaba, 2021). Creatures confined to Earth have little to fear from cosmic rays. Airline crews receive among the most radiation from cosmic rays and even their annual dose is typically well below any region of concern. Otherwise, cosmic rays affect infrastructure, such as power grids, during solar storms, with impacts similar to solar storms.

Living with Outer Space Hazards

Listening to Messages

Artificial satellites orbiting Earth provide near-space observations rich with data on the hazards and vulnerabilities described in this and other chapters as well as on the resources vital to human livelihoods and on the consequences of our abuse of nature. One example mentioned earlier in this chapter was the discovery of a hole in the ozone layer over Antarctica followed by monitoring its repair. A photograph from a spacecraft rounding the back side of the Moon showed the Earth "rising": "Earthrise", a photo taken by the crew of Apollo 8 in 1968, shows the Earth in partial shadow cast by the moon.

The more famous "Blue Marble" photo was snapped in 1972 by Apollo 17's astronauts. This shows the Earth as a small blue jewel in a sea of blackness, spurring on measures to protect the Earth's biosphere. In the Earthly historical context of the then US-USSR "space race", our planet appears without artificially-created and enforced borders. The photo became a symbol of unity and harmony. This view of Earth also supported the overturning of mapping conventions that had been in place for centuries (Cosgrove, 2003). The size of Africa was obvious, rather than being relegated to being smaller than Kalaallit Nunaat (Greenland) as shown on many map projections, and no longer would Europe or North America be the main reference point.

The information or "messages" from space include long-term datasets that assist our understanding of trends and variabilities of hazards and also the location, extent, and condition of resources that support human and non-human life (Chapter 4). Examples are global atmospheric and sea surface temperatures, the spread or contraction of human settlements with and without electricity, ground deformation around volcanoes, and large-scale logging or the burning forests and many other land uses and patterns of human activity. These observations provide immediate, real-time data for determining possible threats and opportunities regarding hazards and vulnerabilities. Examples are volcanic ash plumes, blocks in traffic, breaches in flood walls, water deficits in crops and pastures, and signs of crop failure. When used appropriately, the data and the abilities for instant sharing and communications make us safer, contributing to all disaster-related activities.

This work involves global and local cooperation. The International Charter "Space and Major Disasters" (Bessis et al., 2004) has the tagline "Satellite data to support disaster response worldwide". Organizations register to become authorized users and then have the option to request data from other members when a disaster occurs. The initial charter was signed in 2000 and, across 131 countries, reached its 800th activation on 14 February 2023. Seventeen charter members offer a network of 270 satellites. As a local example, remote observations of sea ice feed into decisions by Inuit in Arctic Canada about whether or not it is safe to go hunting (Landy et al., 2022).

Weapons in Space: Hazard Avoided for How Long?

The Committee on the Exploration and Peaceful Uses of Outer Spaces (COPUOS) is a UN body that seeks to foster international cooperation in the peaceful uses of outer space (COPUOS, 2023). For decades, the US had agreed with the mission of COPUOS and supported it, just as the US had supported the goal of the Antarctic Treaty System to ensure that the Antarctic should not become militarized. The Antarctic Treaty System covers all areas south of 60 degrees south in latitude and has stood since 1959, demonstrating environmental diplomacy that can succeed, as with the Montreal Protocol (UN, 1987). Many seek parallels between the peaceful use of Antarctica and the peaceful use of outer space, including applying the resource extraction components of the Antarctic Treaty System to the Moon, asteroids, and other planets.

In June 2018, then US President Donald J. Trump called upon the US Department of Defense to create a new Space Force as a separate sixth branch of the military that would replace the smaller, limited Space Command created in 1996. This new, expanded, and separate branch was duly formed, and suddenly, the door was open to international competition in weaponizing outer space (Stone & Kuznik, 2019).

Simultaneously, the posture of the US regarding the peaceful uses of outer space had been changing for a while. In 2000, the US had abstained in the 163-0 vote in the United Nations General Assembly on a resolution covering the Prevention of an Outer Space Arms Race (UN, 2000). In 2006, when the UN again approved that resolution by 166-1, the US was the sole opposing vote. In 2001, the US Government's Commission to Assess National Security Space Management and Organization (US Government, 2001) had warned that the US could face a "Space Pearl Harbor" if it did not dominate space, and it recommended that the US military "ensure that the president will have the option to deploy weapons in space" (Stone & Kuznik, 2019, p. 542). The public does not know whether or not there are already weapons in space. The institutional arrangements and infrastructure now exist if a US president or the leadership of another space-faring country decides to pursue the weaponization pathway.

Technological Dependency, Risk, and Vulnerability

For humanity, technology is essential for living with hazards beyond the Earth. Too much might lead to difficulties. Escalating dependence on technology can ramp up vulnerability if there is an erosion of skills and products essential to low-tech alternatives. Depending on phone-based GPS instructions leads to declining map-reading skills (Bachmann, 2015) – and declining availability of paper maps. Ground-based surveys produced maps as a snapshot in time, building up an historical archive. Remote observations that are continually updated overwrite the older data unless a specific time slice is captured, tagged, and preserved. Satellites break or can be shut down during solar storms, while their signals can be jammed inadvertently or deliberately, leaving many floundering to find their way in real-time without computers. Human error in the programming or inadequacy of the algorithms leads people astray due to their faith in what they are instructed. GPS is owned and operated by the US government, making much of the world rely on a single country. Other satellite navigation options include BeiDou from China, Galileo from the European Union, and Glonass from Russia, each with its own advantages and limitations, including security and reliance on a particular government.

These efforts to use outer space for monitoring and supportive action on natural hazards, vulnerabilities, and their impacts can themselves present further off-world hazards. Earth orbits are cluttered by the number of satellites being launched along with the failure to clean them up when they are no longer used or usable (see the Box). The advent of satellite-based fast internet access has led to a thirst for more. Private companies seek to launch thousands of small satellites into low-Earth orbit in order to satiate our demand for unlimited bandwidth on our phone, thus creating even more dependence on satnav and real-time information. How long until these signals can reach us without relays in buildings, in caves, and in tunnels – and what satellites will be required to do so? These networks also give power to the owners to turn internet access on and off at will.

Addressing disasters requires balancing the use of the outer space technologies with maintaining skills and alternatives for inevitable occasions when outer space–reliant technologies are not available, when they fail, or when their limits are reached. As always with disasters, the key remains reducing and avoiding the creation of vulnerabilities. There is an irony that much of the world's population has little opportunity to learn about satellite-based observations of Earth and near-Earth space or the hazards and

BOX 8.2 Space Junk

Space junk is the problem of rocket detritus and obsolete satellites remaining in orbit alongside the wreckage of satellites hit by NEOs, all littering near-Earth space and threatening to destroy other functioning satellites. This space clutter inhibits Earth-based observations of the universe, meaning that monitoring for NEOs and other outer space hazards must impose the expense of placing the instruments in space as discussed earlier in relation to Earthguard and some other monitoring systems. When these NEO-tracking satellites are decommissioned, they add to space junk.

Kessler syndrome refers to the situation in which there is so much space junk that collisions lead to debris that knock out other satellites, increasing the debris and leading to more collisions in a domino effect. Parts of outer space around the Earth would be too dangerous for satellites given the high risk of collision with space debris.

The lack of action on addressing space junk is due to disinterest and little coordinated enforcement. There are no regulations or agreements in place that require satellite owners to decommission and retrieve or safely destroy their obsolete satellites. Some do burn up in the atmosphere – also indicating the risk of them not completely disintegrating and yielding a low-probability-high-impact risk to life, infrastructure, and ecosystems.

opportunities revealed, while a minority of humanity has made itself so dependent on these technologies that they are at risk of creating new risks.

Dying from Outer Space Hazards: Extinction Events?

The universe is awe-inspiring. Planet Earth alone is inspiringly diverse and complex with a long way to go to understand even a small fraction of every facet and link in the web of living and non-living components. Huge uncertainties and unknowns remain, particularly due to the expanse of the universe – its scales of space and time far beyond what any human could experience. Across these scales, many outer space phenomena are highly hazardous. They could cause the extinction of life on Earth or even destroy the planet entirely. Many such hazards grant little-to-no warning, apart from us being able to infer their possibility. Consequently, they lack extensive scope for reducing either the vulnerability or the hazard.

Some stars die with massive explosions called supernovae. Others release sudden spurts of deadly gamma-ray radiation (Palmer et al., 2005). The collision of gravitational waves has a theoretical possibility of leading to a black hole causing planetary-wide harm (Pretorius & East, 2018). Currently, we are not aware of any stars in our vicinity with potential for any of these hazards. We do not fully understand all these phenomena or know exactly when, how, or why they occur. In the geologic past, it is possible that mass extinctions were caused or influenced by these or other off-world hazards.

Black holes are common throughout the universe. A black hole sucks in all matter and energy that slips over a certain threshold, never again to interact with the known universe. No documented black holes sit close by the Earth – that is, none that we have observed or possibly could observe. A black hole with the mass of the Earth would be approximately 2 cm across and would be nearly undetectable until it interacts with another body. That body just might be the Earth if such a mini black hole passes close by or thuds into us. A collision could shatter the planet. A close pass could shift the Earth's orbit, with scenarios sending our planet too close to or too far from the Sun to sustain life in its current form. Other than a wandering black hole, planet, or moon, another star could influence the Moon's orbit or knock the Earth out of its solar orbit. Stars move comparatively slowly and are readily detectable if we are watching. With appropriate surveillance, we would give ourselves a long warning time that trouble looms from afar. These wandering stars add a further potentially existential hazard to a growing list as the understanding of the universe expands.

Given this set of hazards, the odds are not in favor of long-term planetary survival. Humanity and all life on planet Earth would most likely be destroyed if our home planet vanishes. Would it be feasible to redress this planet-wide vulnerability? Theoretically, the hazards themselves could be reduced or eliminated. For example, global geoengineering could address the heat from a passing star with some kind of mirror or filter in outer space or by blocking the extra radiation through injecting aerosols into the Earth's upper atmosphere or supporting cloud formation. Such proposals are currently under investigation as engineering solutions for human-caused climate change, although critics note their likelihood of causing more problems than they solve (Strauss & Burns, 2013). One example using computer simulations suggested that geoengineering to cool the planet would likely increase the frequency and intensity of many types of El Niño and La Niña episodes (Malik et al., 2020). Addressing human-caused climate change has many options (Chapter 5) other than the large risks from global geoengineering.

For the other hazards described in this section, short of building another planet, any scenario involving escape pods or survival bunkers could save only a tiny fraction of humanity and other species. Underground shelters could perhaps be constructed to ride out a supernova, gamma-ray burst, or star's heat but would not help in the context of planetary destruction. If the Earth remains more or less intact post-hazard, then humanity could emerge and begin again to populate and build across the planet, provided the climate were amenable and the natural protections from solar radiation remained. We would even have the opportunity to build a more sustainable, equitable, and peaceful world. An existential hazard from beyond the Earth could thus produce long-term improvements. Could we do so if the oceans, lakes, and rivers had boiled away or if all vegetation were charred or cancerous? These scenarios are yet to be fully explored except in the genre of apocalyptic science fiction.

Were the Earth to be destroyed by an outer space hazard, then fully self-contained escape pods would be the only option. Alas, they might not suffice to perpetuate humanity since they require lead-time for people to reach them and for them to blast off. Assuming that vigilance suffices to provide years of advance warning, what might happen? Choosing the survivors to leave the planet would be difficult and likely to spark social and political conflict. Should we hope that either the Moon or Mars remain available for settlement? Or that Venus or Mercury is kicked into an Earth-type orbit and that we have the capability to swiftly engineer this planet into a New Earth? Should the plans be to form a new society residing in space stations drifting around our solar system, to build an artificial planet in Earth orbit, to spend generations seeking another solar system, or a combination? Humanity might survive as a species, although civilizations and societies as we know them today and understand them to have existed for millennia would be completely finished.

Which is inevitable anyway. We currently have warning that, even if planet Earth remains habitable for the next 5 billion years and humanity somehow still exists in a semi-recognizable form (after all, evolution seems likely to eventually change the species *Homo sapiens*), then the Sun's life must end. It will enlarge, engulfing the inner planets, staying at that size for perhaps a billion years before shrinking and cooling. Will humanity have sufficient geoengineering, solar engineering, solar system engineering, or galactic engineering at that point to ward off or survive the Sun's change? Or will humanity be so different and so distant from the present that the Earth will not even remain in ancient legends?

Irrespective, outer space hazards or human actions will end the Earth as it is known today.

Conclusions

The possibility of no-warning, planetary-wide, extinction-level hazards from outer space brings us to the end of Part II of this book, highlighting the extremes which nature tosses our way. These hazards can and should be dealt with by addressing hazards simultaneously with human and non-human vulnerabilities and abilities. The emphasis must remain focused on vulnerabilities (including exposures) and abilities. Human actions in the foreseeable future might end up with mining and even settlements on the Moon, asteroids, and nearby planets. For some time yet, our actions are unlikely to significantly impact the rest of our solar system, let alone throughout and beyond the Milky Way galaxy.

References Cited

Alabdulgader, A., McCraty, R., Atkinson, M., Dobyns, Y., Vainoras, A., Ragulskis, M., & Stolc, V. (2018). Long-term study of heart rate variability responses to changes in the solar and geomagnetic environment. *Scientific Reports, 8*(1), 11–14. https://doi.org/10.1038/s41598-018-20932-x

Bachmann, M. M. (2015). Exploring and addressing functional map illiteracy in community college students. *Papers in Applied Geography, 1*(3), 258–261. https://doi.org/10.1080/23754931.2015.1014674

Bains, W., Petkowski, J. J., & Seager, S. (2023). Venus' atmospheric chemistry and cloud characteristics are compatible with Venusian life. *Astrobiology, 23*(10), 1–45. https://arxiv.org/ftp/arxiv/papers/2306/2306.07358.pdf

Bessis, J., Béquignon, J., & Mahmood, A. (2004). The international charter "space and major disasters" initiative. *Acta Astronautica, 54*(3), 183–190. https://doi.org/10.1016/S0094-5765(02)00297-7

Brobrowsky, P., & Rickman, H. (Eds.). (2007). *Comet/astroid impacts and human society*. Springer.

Brown, P., Spalding, R. E., ReVelle, D. O., Tagliaferri, E., & Worden, S. P. (2002). The flux of small near-earth objects colliding with the earth. *Nature, 420*(6913), 294–296. https://doi.org/10.1038/nature01238

COPUOS (Committee on the Peaceful Uses of Outer Space). (2023). *U.N. office for outer space affairs*. https://www.unoosa.org/oosa/en/ourwork/copuos/index.html

Cosgrove, D. (2003). *Apollo's eye: A cartographic genealogy of the earth in the western imagination*. Johns Hopkins University Press.

Drake, F. D. (1961). Project Ozma. *Physics Today, 14*(4), 40–46.

Drake, F. D. (1965). The radio search for intelligent extraterrestrial life. In G. Mamikunian & M. H. Briggs (Eds.), *Current aspects of exobiology* (pp. 323–345). Pergamon.

EPA (U.S. Environmental Protection Agency). (2022, December 15). *Health effects of UV radiation*. https://www.epa.gov/sunsafety/health-effects-uv-radiation

Farman, J. C., Gardiner, B. G., & Shanklin, J. D. (1985). Large losses of total ozone in Antarctica reveal seasonal ClOx/NOx interaction. *Nature, 315*(6016), 207–210. https://doi.org/10.1038/315207a0

Huang, S.-S. (1959a). Occurrence of life in the universe. *American Scientist, 47*(3), 397–402.

Huang, S.-S. (1959b). The problem of life in the universe and the mode of star formation. *Publications of the Astronomical Society of the Pacific, 71*(422), 421–424.

Landy, J. C., Dawson, G. J., Tsamados, M., Bushuk, M., Stroeve, J. C., Howell, S. E., Krumpen, T., Babb, D. G., Komarov, A. S., Heorton, H. D., Belter, H. J., & Aksenov, Y. (2022). A year-round satellite sea-ice thickness record from CryoSat-2. *Nature, 609*(7927), 517–522. https://doi.org/10.1038/s41586-022-05058-5

Malik, A., Nowack, P. J., Haigh, J. D., Cao, L., Atique, L., & Plancherel, Y. (2020). Tropical Pacific climate variability under solar geoengineering: Impacts on ENSO extremes. *Atmospheric Chemistry and Physics, 20*, 15461–15485. https://doi.org/10.5194/acp-20-15461-2020

Mayor, M., & Queloz, D. (1995). A Jupiter-mass companion to a solar-type star. *Nature, 378*(6555), 355–359. https://doi.org/10.1038/378355a0

Molina, M. J., & Rowland, F. S. (1974). Stratospheric sink for chlorofluoromethanes: Chlorine atom-catalysed destruction of ozone. *Nature, 249*(5460), 810–812. https://doi.org/10.1038/249810a0

Palmer, D. M., Barthelmy, S., Gehrels, N., Kippen, R. M., Cayton, T., Kouveliotou, C., Eichler, D., Wijers, R. A., Woods, P. M., Granot, J., Lyubarsky, Y. E., Barbier, L., Chester, M., Cummings, J., Fenimore, E. E., Finger, M. H., Gaensler, B. M., Hullinger, D., Krimm, H., . . . Tueller, J. (2005). A giant γ-ray flare from the magnetar SGR 1806–20. *Nature, 434*(7037), 1107–1109. https://doi.org/10.1038/nature03525

Popova, O. P., Jenniskens, P., Emel'yanenko, V., Kartashova, A., Biryukov, E., Khaibrakhmanov, S., Shuvalov, V., Rybnov, Y., Dudorov, A., Grokhovsky, V. I., Badyukov, D. D., Yin, Q.-Z., Gural, P. S., Albers, J., Granvik, M., Evers, L. G., Kuiper, J., Kharlamov, V., Solovyov, A., . . . The Chelyabinsk Airburst Consortium. (2013). Chelyabinsk airburst, damage assessment, meteorite recovery, and characterization. *Science, 342*(6162), 1069–1073. https://doi.org/10.1126/science.1242642

Pretorius, F., & East, W. E. (2018). Black hole formation from the collision of plane-fronted gravitational waves. *Physical Review D, 98*, Article 084053. https://doi.org/10.1103/PhysRevD.98.084053

Santos, N. C., Mayor, M., Naef, D., Pepe, F., Queloz, D., Udry, S., & Burnet, M. (2001). The CORALIE survey for southern extra-solar planets. VI. New long period giant planets around HD 28185 and HD 213240. *Astronomy & Astrophysics, 379*, 999–1004. https://www.aanda.org/articles/aa/pdf/2001/45/aah3054.pdf

Schoene, B., Eddy, M. P., Samperton, K. M., Keller, C. B., Keller, G., Adatte, T., & Khadri, R. S. F. (2019). U-Pb constraints on pulsed eruption of the deccan traps across the end-cretaceous mass extinction. *Science, 363*(6429), 862–866. https://doi.org/10.1126/science.aau2422

Slaba, T. C. (2021). Historical reconstruction of astronaut cancer risk: Context for recent solar minima. *Space Weather, 19*(10), e2021SW002851. https://doi.org/10.1029/2021SW002851

Sprain, C. J., Renne, P. R., Vanderkluysen, L., Pande, K., Self, S., & Mittal, T. (2019). The eruptive tempo of deccan volcanism in relation to the cretaceous-paleogene boundary. *Science, 363*(6429), 866–870. https://doi.org/10.1126/science.aav1446

Stone, O., & Kuznik, P. (2019). *The untold history of the United States.* Gallery Books.

Strauss, A. L., & Burns, W. C. G. (2013). *Climate change geoengineering: Philosophical perspectives, legal issues, and governance frameworks.* Cambridge University Press.

Tsurutani, B. T., Zank, G. P., Sterken, V. J., Shibata, K., Nagai, T., Mannucci, A. J., Malaspina, D. M., Lakhina, G. S., Kanekal, S. G., Hosokawa, K., Horne, R. B., Hajra, R., Glassmeier, K.-H., Gaunt, C. T., Chen, P.-F., & Akasofu, S.-I. (2022). Space plasma physics: A review. *IEEE Transactions on Plasma Science, 51*(7), 1595–1655. https://ieeexplore.ieee.org/stamp/stamp.jsp?tp=&arnumber=9961153

UN. (1987). *Montreal protocol on substances that deplete the ozone layer.* United Nations.

UN. (2000). *Fifty-fifth session. First committee. Agenda item 72. Prevention of an arms race in outer space.* United Nations General Assembly.

US Government. (2001). *Report of the commission to assess national security space management and organization.* US Government.

WHO (World Health Organization). (2023). *Health topics. Ultraviolet radiation.* https://www.who.int/health-topics/ultraviolet-radiation

Wolszczan, A., & Frail, D. A. (1992). A planetary system around the millisecond pulsar PSR1257 + 12. *Nature, 355*(6356), 145–147. https://doi.org/10.1038/355145a0

Chapter Nine

Disaster Risk Construction

Disaster risk reduction (DRR) has become a buzzword. The phrase had a freshness once and a challenging content that spoke truth to power in the pursuit of root causes of vulnerability (Wisner, 2020). Perhaps DRR can be refreshed by a critical assessment of language and vocabulary. Even refurbished and redeployed, the term begs an essential question: Where do disaster risks come from? Since disasters are not natural because natural hazards require a social/human baseline to become a disaster (O'Keefe et al., 1976), one needs to ask about disaster risk creation or disaster risk construction1 (both providing the acronym DRC).

This chapter examines how elite or special interests provoke or allow changes in society, nature, and their interactions in ways that create new and emergent risks. This chapter also explores how misuse of power increases vulnerability to hazards and undermines ability to cope with hazards. In addition, this chapter interrogates how assumptions about control over resources, choices, and opportunities may be manipulated by powerful groups in their own interest. Root causes of disaster vulnerability were discussed in Chapter 1. They include inequality, inequity, marginalization, and oppression. To stop disaster risk creation/construction, such root causes must be addressed. This is not easy, but we make a start in Chapter 10.

Risk is accumulated over decades or even centuries (Bull-Kamanga et al., 2003). Risk is transferred and hidden (Etkin, 2011). Risk created in the historical past often has

DOI: 10.4324/9781003292814-9

contemporary reverberations, as seen in colonialism. The spatial organization of the formerly colonized world largely retains the international boundaries established through colonialization. Regional disparities between center and periphery and the economic dominance of one or two large cities persist in these countries.

Definitions of DRC

DRC is a set of processes that generates, increases, or fails to prevent or reduce vulnerabilities and hazards (Lewis & Kelman, 2012). Bankoff and Hilhorst (2022, p. 7) explain, "Disaster risk creation re-emphasises the causes of disaster risks, and, importantly, recognises that disaster risks are created through human interference with natural hazards, the social production of vulnerability, the neglect of response capacities, or a combination of all three". In their discussion of DRC, Wisner and Lavell (2017) address the question of intentionality (pp. 4–5):

> [W]e believe that disaster risk construction generally proceeds unintentionally with a certain automaticity, a function of the way an entire socio-political-economic or as some would put it, socio-environmental system, works routinely from day to day. This may also be called *congruent risk* because it is inseparably linked with the routine functioning of the system . . . We are concerned with a spectrum of intentionality, of specific occasions and processes of decision making visible in the large amount of urban gentrification and restructuring and many large-scale industrial, energy, mineral and other megaprojects that national and international finance capital have driven over the past few decades; [that is] the willful investment in large-scale changes in environment, land use, topography, drainage and the distribution of population.

Approaching DRC from a slightly different angle, Wisner and Gaillard (2009, p. 6) discuss:

> the neglect of neglect . . . On the national scale . . . disasters reflect neglect of vulnerable groups by national economic and political elites and poor governance. Too many disasters are indeed rooted in the non-respect of existing construction codes, the misuse of public funds, or the looting of natural and economic resources to the benefit of the most powerful . . . One could speak of "neglect of neglect" when donors and international institutions turn a blind eye to political neglect of marginal groups of people, dwelling instead on the trappings of "democratic process" under the rubric of "good governance".

Drivers of DRC

From the 1980s onwards, studies have identified components and drivers of DRC. Table 9.1 illustrates some of the processes that influence the location, pace of change, magnitude, and intensity of DRC. The importance of single or combined variables can vary in time and space. It is possible to use and create diverse research questions, approaches, and methods to analyze these factors, such as those proposed by the pressure and release framework (Blaikie et al., 1994; Wisner et al., 2004, 2012) discussed in Chapter 1 and other frameworks reviewed in Chapter 2.

There is consensus among these authors about the impact of population dynamics, poor land use including unchecked urbanization, skewed distribution of access to land and other resources, corruption or poor governance, and environmental degradation. Distortion of economic relations and public finance as well as displacement of people are listed by the five sources. This is a good start. Greater clarity can be achieved if one returns to the definition of DRC and drills down in detail, putting DRC under the lens provided by ordinary language interrogatives: What? How? By whom? When? Where? Why?

TABLE 9.1 Drivers of disaster risk construction

Davis (1987)	Wisner et al. (2004)	Lewis and Kelman (2012)	Wisner et al. (2012)	Oliver-Smith et al. (2016)
– War/civil strike – Population explosion – Urbanization – Deforestation – Inequitable land ownership	– Rapid population change – Rapid urbanization – Arms expenditure – Debt repayment schedules – Deforestation – Decline in soil productivity	– Environmental degradation – Discrimination – Displacement – Self-seeking public expenditure – Denial of access to resources – Corruption – Siphoning of public money	– Rapid population change and transformation – Rapid urbanization – Fluctuations of the world economy market – Ongoing armed conflict – Government debt repayment schedules – Poor governance and corruption – Land grabbing – Deforestation, mining, and overfishing – Decline in soil productivity – Decline of biodiversity	– Population growth – Urban and rural land use patterns and processes – Environmental degradation and ecosystem service depletion – Poverty and income distribution

What? Expanding the Definition of DRC

In Chapter 1, we defined risk as the chance that exposure to hazard will result in harm to a person whose vulnerabilities exceed their capacities for self-protection, those capacities are blocked, and available social protection is deficient to make up for missing individual capacity. We used a summary mnemonic in the form of a pseudo-equation. "M" stands for mitigation of risk (hazard management and/or social protection). "R" and "H" stand for risk and hazard, respectively. "V" for vulnerability and "C" for capacity are represented in a reciprocal relationship. Greater capacity can reduce the impact of or offset vulnerability. Increased vulnerability can overwhelm existing capacity. Thus:

$$R = (H \times V/C) - M$$

Building on those basic definitions and relations – and recalling the caveat that the mnemonic is not meant to produce quantitative measures of risk – the creation or construction of risk can be defined as an action or process that affects any of these elements in a manner that results in increasing harm or the chance of harm. All such perturbations may result in risk construction. A hazard can be made more intense, widespread, or persistent. Improper road construction across an unstable slope could increase landslide hazard. A policy that under-funds public hospitals, forcing people to pay for service at private facilities, will increase vulnerability to disease of low-income people who cannot pay – and, by extension, will increase everyone's chance of getting the hazard when some infected people cannot get treatment and, often, must still go out to get food and to work, thereby spreading the pathogen. Capacities will be blocked when they are not acknowledged and valued or when their practice is forbidden by law. Local knowledge of the local environment and the use of fire to manage forests have been undervalued by outside "experts" and authorities. Where governments have banned burning, the result has sometimes been an explosion in deer tick infestation that transmits Lymes disease to humans (Gallagher et al., 2022). Finally, social protection and risk management efforts by governments may fail in the face of mistrust by the public or through misinformation and disinformation. The rollout of COVID-19 vaccines in some places suffered this fate (Chapter 4).

How and By Whom? Inquiring into Agency

The state apparatus, government authority, or public administration is not a monolithic entity (Lund, 2006). Rather, it is usually composed of organizations that compete for resources and have conflicting roles and agendas. Some public organizations, such as ministries of trade, industry, or regional development, may trigger DRC, while others, such as departments of environmental affairs, public health, emergency management, and civil protection, try to manage the impacts of disasters – with mixed successes and failures. When thinking about the state, it is essential to consider the financial and human resources that national, subnational, and local organizations control, as well as their share of total resources available. For example, the funding typically allocated for environmental protection is less than that allocated for the construction of public infrastructure, as is the case in Brazil (Damacena et al., 2022).

States are influenced by the ideologies discussed in Chapter 3. The ideologies, power dynamics, discourses, and practices that guide the executive, legislative, and judicial branches of government can conflict with one another. Parliamentarians can lobby for agricultural and mining corporations, proposing measures that reduce environmental regulation, social protection, and agrarian reform policies. Heads of government and heads of state can be authoritarian and rule through decrees or can follow the mandates of the legislature. Even then, an executive may try to nudge legislative decisions in favor of a particular contractor or company. In situations where the public and private interests of those representing citizens are merged, the judicial branch of government can try to influence the decisions of the executive and legislative – or itself may be authoritarian or prone to lobbying. Sometimes, groups of people occupying key positions in governments, transnational corporations, and banks catalyze rapid and extensive land use change and transformation of territory, using their economic, political, and symbolic power (Bourdieu, 1991) to frame a convenient truth that hides disaster risk and protects their interests. The judicial system can be a venue for parties to a dispute over the possible harm or benefit of a project, and in some cases, the judiciary can act on its own to clarify the distribution of benefit and harm from investment. For example, the Indian Supreme Court can be proactive in protecting the environment in decisions that go beyond a conventional court's role as adjudicator of disputes (Sharon, 2022).

When disasters happen and are made visible in the media, many actors – CEOs, technicians, public officers, policy makers, scientists, emergency managers, and so many others – may be blamed and can be investigated (Alexander, 2014; Hall, 2011). Knowledge and power are decisive in disasters (Gaillard, 2021), but knowledge must be applied to a specific, visible threat. This involves media and communication skills. Those with knowledge and/or power may fail in communication and application of what they know. Disaster law and jurisprudence are called upon to determine who needs to be blamed for disasters (Lauta, 2016; Samuel et al., 2019). Scientists working in academia or private companies engage in knowledge battles to frame some disasters (such as dam failures and floods), while slow-onset and everyday hazard processes (such as droughts, water insecurity, and slope instability) often remain unnoticed. Box 9.1 provides two examples that illustrate some of the direct and indirect agents and processes that contribute to understanding how a particular instance of DRC occurred.

BOX 9.1 Disaster Risk Constructors and Builders

Ecuador: Controlling Water and Energy – Constructing Disaster Risks

The Dulcepamba basin in central Ecuador covers nearly 500 km^2 and extends from the ~3,200 m elevation highlands of the Andean plateau down to the coastal foothills at ~100 m above sea level. Land use in the sloping downstream portions of the watershed is dominated by agriculture, including a variety of crops.

In 2012, Hidrotambo S.A. began construction of a water intake facility for a hydroelectric plant in the channel of the Dulcepamba River. A storm and resulting flood on 19–20 March 2015 sent water into the agrarian town of San Pablo de Amalí, causing several deaths, loss of property, bank erosion, and damage to the town's only road access. Residents stated that the construction of the hydroelectric power plant directed flood flows toward the town, causing damage and fatalities.

Newmiller et al. (2015) conducted a hydrologic and hydraulic analysis of the catchment. They concluded that damage to the village would not have occurred without the obstruction created by debris blockage of the hydroelectric plant. Moreover, their research results indicated that the hydroelectric water rights awarded to Hidrotambo by contract would allow Hidrotambo to claim all of the flow in the river during part of the year, leaving little or no flow available for downstream irrigators, particularly during the dry summer season, when irrigation demands are highest. Thus, Hidrotambo would seem to have created both short-term catastrophic risk with the loss of lives and long-term risk to livelihoods by artificially imposing seasonal drought.

Hidrotambo S.A. is a partnership secured by Ecuador's Corporation for Energy Research (CIE) and Canadian, Ecuadorian, Spanish, and Swiss investors. The Hidrotambo dam was built in 2016 with national government permission and without respecting the communities' rights to free, prior, and informed consent

(International Rivers, 2022). Local people tried to stop this intentional construction of disaster risk. Social movements of people affected by such dams have tried to stop these DRC processes. Corporations and investors mobilize to protect their investments.

Philippines: Building New Homes – Constructing Disaster Risks

In 2012, in the aftermath of Typhoon Pablo, the government relocated an entire village community to a high-gradient plateau, citing safety concerns. The state provided cash grants to purchase building materials for the construction of new houses, and labor was provided free of charge by family and friends (Siddiqi, 2022). Forest cover was quickly removed from the plateau, and close to 100 brick-and-mortar houses were built. The municipal government – the mayor, social and welfare department, local council representatives, and municipal disaster risk reduction (DRR) officers – encouraged rapid settlement of this new village on the plateau. The government narrative was anchored in the results of scientific assessments carried out by the national hazard mapping agency.

These assessments had classified the original location of the village as "extremely hazardous and unsafe for human settlement". Siddiqi (2022) conducted interviews with the Municipal Disaster Risk Reduction and Management (MDRRM) officer who confirmed that "this plateau was in fact highly vulnerable to landslides" (Siddiqi, 2022, p. 97). The resettlement was actually related to another political agenda (Siddiqi, 2022), so the disaster was used as a "window of opportunity". According to Siddiqi (2022, p. 96), "A hydro company had moved into the area a year after the typhoon to construct a dam on the part of this river, and a new military checkpoint had been established in the abandoned village to protect the equipment and investment of the company, Euro Hydro Power".

Another example of DRC comes from Brazil. In 2015, a dam containing waste from an iron mine in the state of Minas Gerais collapsed and released a huge mudflow that polluted more than 600 km of water courses, killed 19 people, and displaced hundreds of others along the Doce River (United Nations Human Rights Council, 2015) (Figure 9.1). Four years later, another dam owned by the same company located near the town of Brumadinho in Minas Gerais state collapsed (Agência Brasil, 2019), killing 270 people

FIGURE 9.1 Catholic church still damaged almost three years after the Mariana dam break, in Brazil

Source: photo by Victor Marchezini (1 October 2018)

(Azevedo, 2019). The 2015 disaster had mobilized a chain of investors that had unintentionally constructed risk by financing the construction of the dam. After the 2019 Brumadinho dam disaster, some of the investors, including the Church of England Pensions Board, led the creation of the Investor Mining and Tailings Initiative (IMTSI), a coalition of 112 institutional investors that represents US$14 trillion in assets (Church of England, 2022; Piciullo et al., 2022). Mining companies are reluctant to share information related to their tailings storage facilities (Islam & Murakami, 2021). Four years after the 2019 Brumadinho disaster, the IMTSI had not yet begun conducting audits of good practice by mining companies. In anticipation of a surge in extraction of minerals needed for renewable energy technologies, the United Nations Environment Programme (UNEP) and IMTSI finally began to put in place such a system of audits (United Nations Environment Programme, 2023).

Degrees and shades of meaning of intentionality must be considered. There are direct and indirect actors whose decisions and actions contribute to DRC. For instance, banks finance a project, investors own stock in these corporations, or pensioners have retirement funds including such stock. There are CEOs who make decisions and public officers who evaluate risks, safety measures, and impacts or perhaps who receive bribes or threats to ignore them. For some more remotely indirect actors, such as the pensioner, it would take effort and skill for such an older person to request and understand an analysis of their retirement portfolio that would report on investments' "environmental, social, and governance" (ESG) impacts. Not knowing and not knowing how to know are nevertheless forms of unintentional DRC.

In this chain of agents, usually the most critical question is who it is who funds DRC. Megaprojects would not be built and the associated disaster risk would not be constructed if funding were not available. Wisner (2020) noted that the sources of finance for "development" are changing from the dominant post–World War II, Breton Woods centers of power – particularly the International Monetary Fund and World Bank Group – to alternative sources of finance led by China and by economic alliances among and between emerging larger economies, including the BRICS (Brazil, Russia, India, China, and South Africa) countries and oil-producing countries with petrodollars to invest, including the Gulf States, Indonesia, and Nigeria (Wisner, 2020). Moreover, quite apart from nationally or regionally-based sources of finance, there are multinational corporations around the world that are richer than many sovereign states. The 2,000 largest companies in the world in terms of sales, profits, assets, and market value represented US$47.6 trillion in revenues, US$5.0 trillion in profits, US$233.7 trillion in assets, and US$76.5 trillion in market capitalization in 2021 (Murphy & Contreras, 2022). These companies are headquartered in 58 countries, but three countries are home to 57% of the 2,000 companies listed: the USA (590 companies), China/Hong Kong (351), and Japan (196). The top ten richest companies include five banks: three based in China and two in the US.

Mozambique provides an example of the power possessed by large banks to construct risk. One only needs to consider the debt burden under which Mozambique labors and this debt's catastrophic knock-on effects. In 2012 and 2013, international banks connived with corrupt government officials at the highest level in Mozambique to create a debt that was hidden from Mozambique's parliament and overseas donors (Artur, 2022; World Bank, 2022). Much of the money was siphoned away by officials. This debt was one of the factors that constructed disaster risk. Because discovery of the so-called "hidden debt" caused donors to suspend aid flows, Mozambique had to impose fiscal austerity. The result was that local governments were unable to maintain flood control infrastructure and storm drains or to fund health care and other services. This left them highly vulnerable to the impact of flooding during a series of cyclones (Box 9.2.).

BOX 9.2 Banks and Corrupt Officials – Cyclones, Dependency, and Disaster Risk Construction

In 2016, a hidden international debt was revealed. Government officials had requested state-backed loans – guaranteed without parliamentary approval – to create three state-owned enterprises in 2013–2014. They had incurred more than $2 billion of debt which represented 12% of Mozambique's GDP in that period. In 2016, $1.3 billion of it was undisclosed until international media reports (Gebregziabher & Sala, 2022). Since these loans violated the International Monetary Fund's (IMF) and the International Development Association's money borrowing policies (Gebregziabher & Sala, 2022), budget support by both institutions and other development partners was suspended. Mozambique still depends on international aid for 40% of its annual budget. Consequently, the country was forced to implement fiscal measures to gradually reduce this public debt. To achieve these cuts in public expenditure, public services, such as agricultural extension and road and bridge maintenance amongst others, were reduced (Artur, 2022). Rising international prices for key imports, such as fuel and food, exacerbated an already fragile situation (Artur, 2022).

The 2019 riverine flooding in Mozambique occurred in this context. Cyclone Idai had dropped large amounts of rainfall upstream in Malawi and Zambia, leading to a collapse of dams and bridges in those countries and leading to floods downstream in Mozambique. The city of Tete was partly submerged due mainly to dam and bridge collapses in Malawi. This disaster happened in a context where illegal construction of houses, including in the vicinity of river banks, had become typical practice (Artur, 2022). Austerity put in place due to the hidden debt crisis meant that district and municipal governments had not been able to maintain flood protection and urban storm drainage systems. Robust local government action could have reduced the damage done by the flooding, but austerity had weakened their institutions.

Mozambique had still not recovered when a second cyclone named Freddy made landfall near the central city of Beira and brought torrential rainfall to the region once again. Freddy and Idai set records for their size and persistence. Mozambique was experiencing a cholera outbreak at the time, worsened by the flooding. Nearly 200 people died in these floods in Mozambique, with at least 1,200 deaths in Malawi and more fatalities in other countries (Staub, 2019). Food shortages due to flooding of crops and ongoing cholera added misery and contributed to the fragile situation. The highly-indebted government was totally dependent on donors to provide aid for roughly 1.8 million affected people.

Interlude: The Role of Corruption

Corruption was involved in creating the debt burden in Mozambique and constructing the disaster risk that was realized in deaths, illnesses, displacement, destruction, and loss from cyclones and floods. Corruption is defined as the abuse of entrusted power for private gain. Depending on the amount of money lost and the sector of governance in which the loss occurs, corruption can be classified as "grand", "political", and "petty" (Lewis, 2017) (Box 9.3). The Organization for Economic Co-operation and Development (OECD) states that one-half of bribes paid are in industries in sectors with the largest spending on infrastructure, namely, extractive (19%), construction (15%), and transportation (15%) (Organization for Economic Co-operation and Development, 2014). Poorly cited and poorly built housing and infrastructure exacerbates hazards and increases vulnerability.

BOX 9.3 Types of Corruption

Grand corruption in governments usually involves politicians and commercial operators who are known to have collectively siphoned vast amounts of money, much of it from development funding, often from their own disaster-prone countries and often into private bank accounts in the countries that were the aid's origin (Ndikumana & Boyce, 2011).

Political corruption is the manipulation of policies, institutions, and rules of procedure in the allocation of resources and financing and procurement protocols by decision-makers, who use their position to sustain their power, status, and wealth and to benefit their cronies (Transparency International, 2016). Political corruption may occur when specific providers of goods and services to governments, including construction firms that bribe officials to approve substandard goods or services, such as a poorly constructed building.

Petty corruption refers to the everyday abuse of entrusted power by low- and mid-level officials in their interactions with ordinary citizens when they are looking for access to services and basic goods (Transparency International, 2016; Lewis, 2017).

Corruption contributes to DRC because it undermines compliance with building regulations and diverts investments that should support social protection policies and DRR (Lewis & Kelman, 2012). Lack of local government funds to maintain storm drains before Cyclones Idai and Freddy made landfall in Mozambique is a good example of the impact of grand corruption. Bribes paid to building inspectors in Türkiye to approve poor work and resulting building collapses is an example of political corruption. Sanderson et al. (2022, p. 3) state about the effects of corruption.

> The effects of corruption on the location, design, construction, and maintenance of buildings and infrastructure result in dangerous built environments, which kill, maim, and make large numbers of people homeless every year. Those living in poor-quality buildings and using poor-quality infrastructure are often those more at risk. After all, when disasters occur, it is building and infrastructure collapses that claim the lives of people.

Ambraseys and Bilham (2011) identified that over a 30-year period, about 83% of all deaths attributable to earthquake-related building collapse happened in countries that were considered "anomalously corrupt". There are many examples worldwide showing the relationships between corruption and disasters (Lewis, 2017; Sanderson et al., 2022). Corruption in the form of misappropriation of funds has been traced to deficient maintenance of water and sanitation infrastructure and resulting outbreak of cholera, leading to deaths, in South Africa and adjacent southern African countries (du Plessis, 2023).

Given the tight connections among social, economic, and political structures, it is easy to see in the triangle of vulnerability (Figure 9.2) that corruption in political, economic, and social structures can reinforce each other and contribute to locking some people into unsafe locations and fragile livelihoods by reinforcing the processes of marginalization. The result is DRC as a byproduct of the pursuit of special interest.

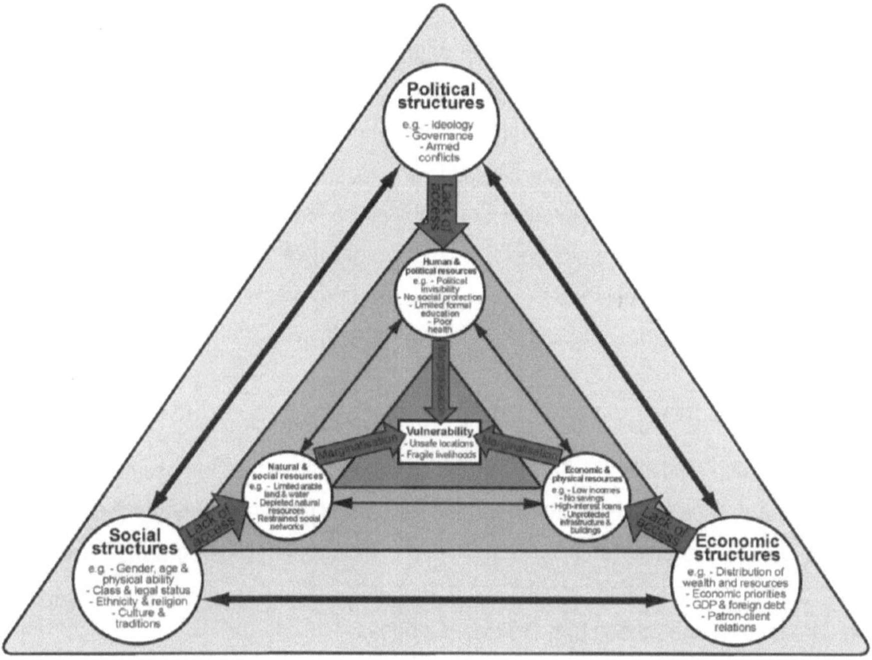

FIGURE 9.2 Triangle of vulnerability

When? Where? Deep Diving into Specifics

Historical and geographical scales are crucial to understanding DRC processes. "When" and "where" can help us to reflect on "how" and "by whom" – and eventually on "why". Colonialism, slavery, and other forms of domination reach out of our shared past and affect us more than we are usually aware. Some predisposing vulnerabilities may have been created over deep historical time measured in centuries or longer. DRC is often a consequence of past coloniality as Quijano (2000, p. 381) defines it, "a structure of domination/exploitation, where the control of the political authority, means of production, and labor of a given population is exercised by another population who claims a different identity and whose government is located in another territorial jurisdiction". European powers' quest for natural resources and cheap labor required the construction of disaster risk in remote places they dominated throughout their colonial period. Since formal decolonization, DRC has continued in those formerly colonized territories through the hegemonic rule of capitalism and neoliberalism. Bonilla (2020) has shown how this matrix of colonial and decentralized domination has been central to the construction of disaster risk in Puerto Rico, leading to the catastrophic impact of Hurricane Maria in 2017.

Colonial and neocolonial economic and political domination have been legitimized by a binary understanding of the world in which the (former) colonies are regarded as "retarded" or "traditional" in comparison to "advanced" and "developed" Europe and its offshore settler territories (Bankoff, 2001). This discourse of otherness/alterity and retardment led to the imposition of Western science and technology as the best way to prevent harmful impacts of natural phenomena. Culturally-embedded and long-tested local strategies for hazard prevention were either ignored or dismissed as primitive. The effect was to reinforce dependence on external Western expertise and to erode local people's abilities/capacities as, for example, in the case of Yungay, Peru (Oliver-Smith, 1994).

Temporal Dimensions of DRC

The retrospective analysis of DRC processes can use various temporal scales – months, seasons, years, decades, centuries, or longer. For instance, when analyzing the causes of the 1970 earthquake disaster in Yungay, Peru, just mentioned, Oliver-Smith (1994) took a centuries-long view that had brought about increased population density in earthquake-prone areas and construction of more fragile houses.

There are centuries-old examples of people questioning the causes of disaster risk construction. One example is the dialogue between Voltaire and Rousseau about the 1755 earthquake in Lisbon, Portugal (Dynes, 2000). There were other discussions about less famous disasters. For instance, after the 1882 flood in Veneto, Italy:

> there was a heated debate on whether the government was to blame in three respects: the solidity of dikes and the river infrastructure, for which the local and provincial authorities were responsible; the deforestation that had led to increased gully erosion and thus to landslides; and finally the organization of the rescue effort.
>
> (Brice, 2002, pp. 65–66)

Marchezini and Wisner (2017) utilize the progression of vulnerability framework to discuss a decades-long cascade of root causes and dynamic pressures that eventually led to unsafe conditions for people living in flood-prone regions of greater Manaus, in the Brazilian Amazon basin. The authors specify urbanization and rapid population change during the three decades after the creation of the free-trade zone in Manaus during the military dictatorship as can be seen in the two maps in Figure 9.3. In 1970, Manaus had 263 factories and 300,000 inhabitants. In 2000, there were 1,405 units and almost 1.4 million people in the city (IBGE, 2017; IPEA, 2017). In 1999, the Amazon river could

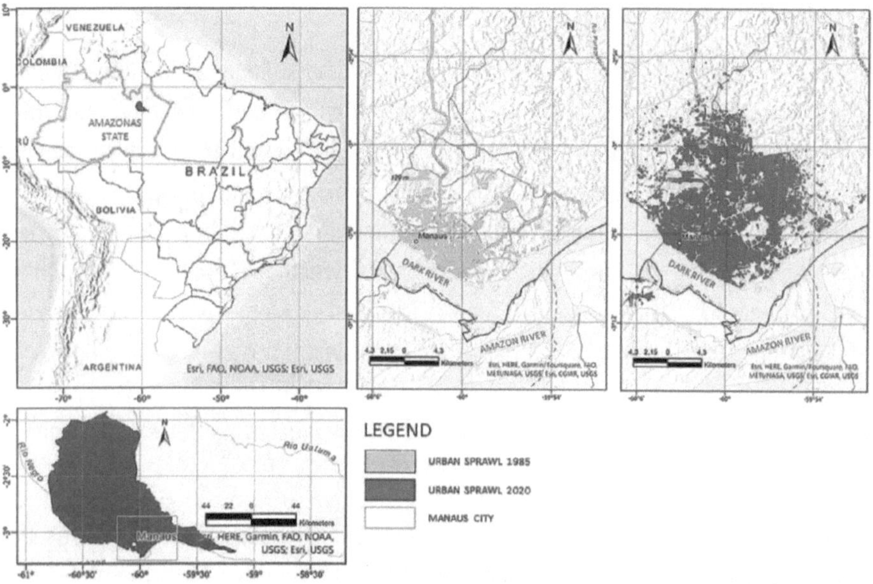

FIGURE 9.3 Poor urbanization and other land use changes (gray areas) around Manaus City in July 1973 (on the left) and August 2007 (on the right)

achieve a height of 29.30 m above sea level and affect 42,000 inhabitants of peri-urban wetlands – people locally known as *igarapés* (CPRM - Serviço Geológico do Brasil, 1999). Ten years later, the peak river level was 0.47 m higher (29.77 m) and could inundate more than 400,000 inhabitants. Seasonality of the hazard inputting into DRC overlaps with the temporalities of other processes that influence DRC.

The pace of DRC processes can vary. Some zones can face rapid DRC pulses influenced by rapid changes in economic investment and disinvestment, sudden increase or loss of population due to migration and displacement, or changes in the permeability of international boundaries. Mega-events are examples of this process. According to the United Nations Human Rights Council (2011, p. 8), during the preparation for the 1996 Olympic Games in Atlanta, "1,200 social housing units for the poor were destroyed in preparation for the Olympic Games. In Sydney [in 2000], reports suggest that around 6,000 people were made homeless in the run-up to the Olympics". In Brazil, 170,000 were displaced due to mega-events, including the 2014 World Cup and the 2016 Olympic Games (Ancop, 2012). According to Flyvbjerg (2014), Greece's 2011 debt default was the result of the cost of staging the 2004 Olympic Games in Athens, another mega-event "for which cost overruns and incurred debt were so large they negatively affected the credit rating of the whole nation, substantially weakening the economy in the years before the 2008 international financial crisis" (p. 16).

Spatial Dimensions of DRC

Construction of risk affects some places more than others. Impacts and consequences depend on scale and preexisting relations between human and non-human life. Rural areas worldwide facing intense DRC processes, mainly related to dynamic pressures, such as new or increased hazards, such as avian flu and other viruses (Chapter 4), population dynamics, land use change, environmental degradation, changes in access to land and security of tenure, and changes in the weather (Chapter 5). Geopolitical power and investment decisions are root causes that drive these dynamic pressures. The numerous individual human beings in chains of institutions, including banks, pension and investment funds, and governments, may not be aware of the consequences of such investment decisions or may not care. Should we understand this as intentional or unintentional DRC?

For instance, about 90 million hectares of arable land have been purchased or leased by foreign investors in sub-Saharan Africa, Asia, and Latin America since the early 2000s (Müller et al., 2021). After the food crisis and increased energy prices in 2007–2008, large-scale investments in land increased, and they established commercial crop production in response to global demand for food and biofuel (Yang & He, 2021). The impact on small farmers, herders, forest workers, and others in these regions can be intense and negative. Such dynamic pressures have exacerbated inequality, inequity, and vulnerabilities to disaster, especially for the poor and landless. Land access inequality and inequity have reached extreme levels in some parts of the world. South Asia and Latin America exhibit the worst with the top 10% of landowners capturing up to 75% of agricultural land, followed by Africa at around 55–60% (Bauluz et al., 2020).

DRC may take place in various ecosystems and biomes, revealing a further spatial dimension. In Shanghai, China, rapid urbanization has triggered unsafe livelihoods for people, with biodiversity and ecosystem loss in formerly rural, now peri-urban areas, due to conversion of wetland islands for industrial use (Siciliano, 2014). In Bangladesh, delta wetlands have suffered encroachment by the expanding urban area, changing waterways and leading to flooding. The development of housing and industry on the former wetlands also released wastewater into rivers without purification, leading to poor-quality water downstream (Feldman & Geisler, 2012).

These are examples of what one might call "horizontal" DRC, where large portions of land are affected. But there are situations of "vertical" DRC. Such risk may not be represented on risk maps because it is lodged in multi-story buildings. Flyvbjerg (2014) recalled that in 1930, New York's Chrysler Building was the tallest building in the world, at 319 m. Dubai's Burj Khalifa building now reaches 828 m – an increase of more than 100% in building height around 80 years. Workers and residents in such megastructures are at risk from structural fire and air circulation hazards, such as Legionnaires' disease. Yet urbanization might not be inherently DRC. More people living and working in dense urban areas can release surrounding surface areas for agriculture and density creates other advantages for some social groups (Kelman, 2016). More people in a place may have easier access to health care facilities, emergency services, paved roads, and other infrastructure and services. High-density living can nonetheless mean that more people are affected by a particular hazard. Many of these forms of vertical DRC are legal, but there are also illegal forms, such as the risk literally "constructed" with poor-quality materials and methods.

Mexico City provides an example of vertical DRC. The historical city grew up on deposits of an ancient lake. The sediments amplify seismic waves produced along the Pacific Coast by offshore earthquakes. This produces ground shaking in what has become an area with high population density. Consequently, particular attention must be given to building construction. The impact of the earthquakes in 1985, 2017, and 2022 revealed corruption in the real estate sector involving economic losses and fatalities. Some of the buildings that collapsed during the 2017 earthquake were built before 1985 and, thus, before the establishment of the 1987 building code. One might consider those responsible for pre-code construction to have been involved in unintentional DRC, if their motivation might have been to provide a public good in the form of housing. A considerable percentage of the collapsed or damaged buildings contained irregularities in their construction (Galvis et al., 2017). Did the officials, bankers, investors, architects, engineers, construction company executives, and others in the chain that led to that risky situation commit intentional DRC? According to de Luna Jiménez and Gómez García (2019), the complex and diffuse legal framework of the real estate sector in Mexico encourages corruption and dilutes responsibilities. This situation makes accountability difficult and fosters impunity.

Why? Purposes, Goals, Motivations, and Intentions

"Why" is a deceptively simple little word in English and many other languages. It may call forth a functional response: to serve this or that purpose. Purpose can connote an immediate or short-term use or function. The purpose of applying nitrogen, phosphorus, and potassium to soil is to increase fertility. The time horizon implied is the cropping cycle. Goals are more strategic and work on a longer time scale. Purposes and goals may overlap. The chemicals added to the soil will have a residual effect on future crop cycles. Nevertheless, the word "why" may interrogate goals. By asking why a government prioritizes education or primary health care, one is seeking to know the longer-term outcome the government is seeking.

One can inquire about motivations. Motivations are often multiple and mixed. For the question "Why did the city mayor issue an evacuation order before the cyclone's landfall?" many answers are possible. The mayor may have feared blame if city residents were injured or killed by storm surge, wind, or freshwater flooding, even though in many cases blame is not equated to legal responsibility. Or perhaps the mayor was acting in accordance with standard procedures upon receipt of advice from "experts". The mayor may have been trying to model leadership and control with an eye on future political advancement. No matter what purposes, goals, and motivations influence such a decision, once made and action taken, there will be intended and unintended consequences. Some consequences involve construction of disaster risk. There may be deaths from vehicle crashes during the evacuation, due to poor planning or poor driving, both of which are easily preventable. DRC may also result from insufficient knowledge of the hazard. The cyclone's size, speed, or trajectory may be misjudged, and the timing of evacuation may be wrong as a result. If the mayor accepted the information they were given, and perhaps better information could not have been available, then DRC could be unintentional.

One assumes the mayor in this hypothetical example is operating in peacetime. During wars or in the midst of civil conflicts, people may suffer intentional DRC. Hazards may be manipulated as weapons. In 1672, during a war with France, parts of the Netherlands were deliberately flooded to deter invading soldiers, a strategy repeated during the Second World War (Wagret, 1968). Another example is provided by Wisner (1993, p. 134), who stated that the US bombed dikes in North Vietnam in a failed attempt "to drown and starve civilian populations of an enemy into submission". More recently, the huge Kakhovka dam on the Dnipro River in southern Ukraine was destroyed by an internal explosion most likely detonated by Russian forces that controlled the dam (Lakezina, 2023). Flooding downstream was widespread.

Warfare and preparation for war may unintentionally construct disaster risk. In Armenia, standardized, modular, cheap apartment buildings collapsed during the 1988 earthquake. The reason for collapse was partly diversion of resources in the Soviet Union from such civilian construction projects into defense spending at the height of the Cold War (Wisner, 2003). Wars may undermine the ability of central or local governments to mitigate hazards and tackle vulnerability, as well as undermining the trust between government and citizens that is essential for prevention and mitigation (Waldmüller & Robledo, 2021; Wisner et al., 2004).

Megaprojects

Megaprojects "are large-scale, complex ventures that typically cost US$1 billion or more, take many years to develop and build, involve multiple public and private entities, are transformational, and impact millions of people" (Flyvbjerg, 2014, p. 6). Although, when first critiqued, megaprojects, such as dams in Africa, displaced "only" tens of thousands, not "millions", and were less costly than US$1 billion (e.g., Scudder et al., 1993). Such large-scale modifications of the landscape hold great potential for unintentional DRC – yet the consequences are well-known and well-documented, so perhaps their construction is really intentional DRC. Large dams illustrate the positive and negative aspects of megaprojects.

The International Commission on Large Dams (ICOLD) defines large dams as having a height of 15 m or greater or between 5 m and 15 m impounding more than 3 million cubic m (International Commission on Large Dams, 2022). In April 2020, ICOLD (2022) recorded the existence of 58,713 large dams worldwide. In 39,110 cases, it was possible to identify their purpose. Irrigation (47%), hydropower (21%), and flood control (12%) were the three main single-purposes. Ninety-five countries provided information about large dams to ICOLD, but only ten countries accounted for 84% of the projects. The top four were China (23,841), the US (9,263), India (4,407), and Japan (3,130). Although dam construction often displaces people, ICOLD did not provide the number of resettled persons for all the large dams. In an analysis of ~7,000 of the largest dams worldwide covering 120 rivers in at least 70 different countries, Richter et al. (2010) estimated that 472 million people were living in rural areas downstream of the dams who had suffered direct and indirect impacts.

Dam failure is a disaster risk. Bernard-Garcia and Mahdi (2020) identified 3,861 cases of historical dam failures around the world based on a literature review of 196 publications. Some of these disasters involved active or inactive mine waste (tailings) dams, such as the Brumadinho dam presented earlier. Human deaths and cascading environmental impacts are common results of dam failure (Islam & Murakami, 2021; Piciullo et al., 2022). A total of 366 tailing dam disasters were identified between 1915 and 2020. Most tailing dam failures were associated with copper mining (86 cases), followed by extraction of gold (66), coal (28), phosphorus (26), lead (21), iron (19), uranium (16), and silver (11). Fifty-four disasters accounted for 2,976 human lives directly lost (Islam & Murakami, 2021; Piciullo et al., 2022).

Some mine tailing dam disasters were not reported, especially those that happened before the 2010s and in countries with lax environmental regulations and a lack of media freedom (Islam & Murakami, 2021). The number of reported disasters was lower in the 2000s compared to that of previous decades, but the records in the 2010s showed a rising trend, which could be linked to the worldwide demand for the minerals being mined (Venter et al., 2016; Cano et al., 2020). For instance, the production of lithium-ion batteries by 2050 will require the construction of 80 Brumadinho-scale tailing dams (Kosai et al., 2020).

The monetary value used in defining megaprojects is arbitrary and easily contested. Irrespective, "mega" projects are increasing in the size of intervention, becoming "giga" and "tera" in scale (Flyvbjerg, 2014). China's Belt and Road Program illustrates this trend (Box 9.4).

This and other megaproject developments are driven by political, economic, technological, and aesthetic motivation (Flyvbjerg, 2014). As with all categories, they and their interpretation have limitations, yet they provide a useful starting point for thinking about megaprojects.

BOX 9.4 China's Belt and Road Initiative

The Belt and Road Initiative (BRI) was proposed at the Bo'ao Forum in 2014 by Chinese President Xi Jinping. This project is a network of highways, rail routes, marine routes, oil and natural gas pipelines, and other infrastructure that involve three continents, eight regions, and more than 65 countries with 4.4 billion residents (Peng et al., 2017). China allocated US$40 billion to the Silk Road Fund and US$100billion to the Asian Infrastructure Investment Bank (AIIB), a multinational financial institution. The budget estimate to implement the BRI project is much larger – US$1.7 trillion annually until 2030 (Enderwick, 2018).

The actors who construct risk by developing megaprojects are usually so focused on the success of their venture and its economic and other "sublime" rewards that they ignore the project's risk implications. This is often true of megaproject investments that displace people from their territories – for example, large-scale agribusiness, hydropower, mining, new technology cities, stadiums, luxury housing development, energy extraction, new ports, and satellite cities (Flyvbjerg, 2014; Lewis & Kelman, 2012; Wisner, 2020).

Summary and Conclusions

This chapter asked questions about root causes and dynamic pressures that contribute to disaster risk construction using five questions: How? By who? When? Where? And why? It also discussed some forms of intentional and unintentional DRC, using historical and contemporary examples from different parts of the world. DRC can result from hazard manipulation, decrease or blockage of people's capacity to cope with risk, increase in vulnerability, and/or reduction of risk mitigation policies and practices by governments. Megaprojects, in particular, are likely to construct risk for the people displaced, for those who remain, for people who are attracted by employment opportunities, and for people living elsewhere at some distance where risk may have been transferred. An example is those who live in one of China's Belt and Road Initiative corridors. The growth of worldwide demand for energy, food, freshwater, fiber, and minerals combines with population growth, increased consumption per person, and poor land use, including unchecked urbanization, to produce a situation in which it is difficult to reduce risks currently being constructed and to address the backlog of accumulated risk.

Note

1 The word "construction" follows usage in the rich literature that has come from the network of social scientists in Latin America who study disaster, La Red de Estudios Sociales en Prevención de Desastres en America Latina. https://www.desenredando.org/

References Cited

Agência Brasil. (2019). *Desastre em Brumadinho*. Agência Brasil. https://agenciabrasil.ebc.com.br/sites/default/files/thumbnails/image/pr_abr_26011910854.jpg

Alexander, D. (2014). Communicating earthquake risk to the public: The trial of the "L'Aquila Seven". *Natural Hazards, 72*, 1159–1173.

Ambraseys, N., & Bilham, R. (2011). Corruption kills. *Nature, 469*(7329), 153–155. https://doi.org/10.1038/469153a

Ancop. (2012). *Mega-events and human rights violations in Brazil.* Ancop – National Network of Popular World Cup Committees. https://www.fundobrasil.org.br/en/projeto/national-network-of-popular-world-cup-committees-ancop-sao-paulo/

Artur, L. (2022). Vulnerability and resilience in a complex and chaotic context: Evidence from Mozambique. In G. Bankoff & D. Hilhorst (Eds.), *Why vulnerability still matters: The politics of disaster risk creation* (pp. 143–167). Routledge.

Azevedo, R. (2019, January 25). Ações da Vale acumulam alta de 258% desde desastre de Mariana. *Valor.* https://valor.globo.com/empresas/noticia/2019/01/25/acoes-da-vale-acumulam-alta-de-258-desde-desastre-de-mariana.ghtml

Bankoff, G. (2001). Rendering the world unsafe: 'Vulnerability' as a Western discourse. *Disasters, 25*(1), 19–35.

Bankoff, G., & Hilhorst, D. (2022). Introduction. In I. G. Bankoff & D. Hilhorst (Eds.), *Why vulnerability still matters: The politics of disaster risk creation* (pp. 1–12). Routledge.

Bauluz, L., Govind, Y., & Novokment, F. (2020). *Global land inequality. WID.world working paper no. 2020/10.* https://wid.world/document/global-land-inequality-world-inequality-lab-wp-2020-10/#:~:text=According%20to%20our%20benchmark%20metric%20of%20agricultural%20land%20inequality%20

Bernard-Garcia, M., & Mahdi, T. (2020). A worldwide historical dam failure's database. *Borealis.* https://doi.org/10.5683/SP2/E7Z09B

Blaikie, P., Cannon, T., Davis, I., & Wisner, B. (1994). *At risk: Natural hazards, people's vulnerability and disasters.* Routledge.

Bonilla, Y. (2020). The coloniality of disaster: Race, empire, and the temporal logics of emergency in Puerto Rico. *Political Geography, 78.* https://doi.org/10.1016/j.polgeo.2020.102181

Bourdieu, P. (1991). *Language and symbolic power.* Polity Press.

Brice, C. (2002). 'The king was pale': Italy's national-popular monarchy and the construction of disasters, 1882–1885. In J. Dickie, J. Foot, & F. Snowden (Eds.), *Disastro! Disasters in Italy since 1860: Culture, politics, society* (pp. 61–79). Palgrave.

Bull-Kamanga, L., Diagne, K., Lavell, A., Leon, E., Lerise, F., MacGregor, H., Maskrey, A., Meshack, M., Pelling, M., Reid, H., Satterthwaite, D., Songsore, J., Westgate, K., & Yitambe, A. (2003). From everyday hazards to disasters: The accumulation of risk in urban areas. *Environment and Urbanization, 15*(1), 193–203.

Cano, N., Hasenstab, C., & Velasquez, H. (2020). Energy life cycle assessment indicators in Colombian gold mining sector. *Journal of Sustainable Mining, 19*(3), 150–165. https://doi.org/10.46873/2300-3960.1013

Church of England. (2022). *The investor mining and tailings safety initiative.* Church of England. https://www.churchofengland.org/about/leadership-and-governance/church-england-pensions-board/pensions-board-investments/investor-1

CPRM - Serviço Geológico do Brasil. (1999). *A Cheia de 1999 em Manaus*. CPRM - Serviço Geológico do Brasil. http://www.cprm.gov.br/sace/rehi/manaus/cheia

Damacena, F., Pereira, L., Costa, R., & Marchezini, V. (2022). Fundos públicos federais e implementação da política nacional de proteção e defesa civil no Brasil. *Revista de Informação Legislativa, 59*, 215–242.

Davis, I. (1987). Safe shelter within unsafe cities: Disaster vulnerability and rapid urbanisation. *Open House International, 12*(3), 5–15.

de Luna Jiménez, J., & Gómez García, G. (2019). *Diagnóstico de corrupción en el sector inmobiliario*. Mexicanos Contra la Corrupción y la Impunidad.

du Plessis, A. (2023, May 23). Cholera in South Africa: A symptom of two decades of continued sewage pollution and neglect. *The Conversation*. https://theconversation.com/cholera-in-south-africa-a-symptom-of-two-decades-of-continued-sewage-pollution-and-neglect-206141.

Dynes, R. (2000). The dialogue between Voltaire and Rousseau on the Lisbon earthquake: The emergence of a social science view. *International Journal of Mass Emergencies and Disasters, 18*(1), 97–115.

Enderwick, P. (2018). The economic growth and development effects of China's one belt, one road initiative. *Strategic Change, 27*, 447–454. https://doi.org/10.1002/jsc.2229

Etkin, D. (2011). Risk transference and related trends: Driving forces towards more mega-disasters. *Global Environmental Change Part B: Environmental Hazards, 1*(2), 69–75.

Feldman, S., & Geisler, C. (2012). Land expropriation and displacement in Bangladesh. *Journal Peasant Studies, 39*, 971–993. https://doi.org/10.1080/03066150.2012.661719

Flyvbjerg, B. (2014). What you should know about megaprojects and why: An overview. *Project Management Journal, 45*(2), 6–19. https://doi.org/10.1002/pmj.21409

Gaillard, JC. (2021). *The invention of disaster: Power and knowledge in discourses on hazards and vulnerability*. Routledge.

Gallagher, M., Kreye, J., Machtinger, E., Everland, A., Schmidt, N., & Skowronsk, N. (2022). Can restoration of fire-dependent ecosystems reduce ticks and tick-borne disease prevalence in the eastern United States? *Ecological Applications, 32*(7), e2637. http://doi.org/10.1002/eap.2637

Galvis, F., Miranda, E., Heresi, P., Dávalos, H., & Silos, J. (2017). *Preliminary statistics of collapsed buildings in Mexico City in the September 19, 2017, Puebla Morelos earthquake. Technical report*. John A. Blume Earthquake Engineering Center, Stanford University.

Gebregziabher, F., & Sala, A. (2022). Mozambique's 'hidden debts': Turning a crisis into an opportunity for reform. *World Bank Blogs*. https://blogs.worldbank.org/africacan/mozambiques-hidden-debts-turning-crisis-opportunity-reform

Hall, S. (2011). Scientists on trial: At fault? *Nature, 477*, 264–269. https://doi.org/10.1038/477264a

IBGE. (2017). *Banco de Dados Agregados*. https://sidra.ibge.gov.br/tabela/1287#resultado>.

ICOLD – International Commission on Large Dams. (2022). *General synthesis*. https://www.icold-cigb.org/GB/world_register/general_synthesis.asp.

International Commission on Large Dams. (2022). *World register of dams: General synthesis*. International Commission on Large Dams.

International Rivers. (2022). *A ticking time bomb: Tomorrow's hearing can decide the fate of companies responsible for dangerously built dam that threatens more than 3,400 families in Ecuador*. International Rivers. https://www.internationalrivers.org/news/a-ticking-time-bomb-tomorrows-hearing-can-decide-the-fate-of-companies-responsible-for-dangerously-built-dam-that-threatens-more-than-3400-families-in-ecuador/

IPEA. (2017). *Ipeadata*. http://www.ipeadata.gov.br/Default.aspx

Islam, K., & Murakami, S. (2021). Global-scale impact analysis of mine tailings dam failures: 1915–2020. *Global Environmental Change*, *70*, 102361. https://doi.org/10.1016/j.gloenvcha.2021.102361

Kelman, I. (2016). Disaster vulnerability by demographics? *The Journal of Population and Sustainability*, *4*(2), 17–30. https://doi.org/10.3197/jps.2020.4.2.17

Kosai, S., Takata, U., & Yamasue, E. (2020). Natural resource use of a traction lithium-ion battery production based on land disturbances through mining activities. *Journal of Cleaner Production*, *280*, 124871. https://doi.org/10.1016/j.jclepro.2020.124871

Lakezina, V. (2023, June 6). Villagers flee floods as huge Kakhovka dam destroyed in Ukraine war zone. *Reuters*. https://www.reuters.com/world/europe/russia-says-it-thwarts-another-major-ukrainian-offensive-donetsk-2023-06-05/

Lauta, K. (2016). *Disaster law*. Routledge.

Lewis, J. (2017). Social impacts of corruption upon community resilience and poverty. *Jàmbá: Journal of Disaster Risk Studies*, *9*(1). https://doi.org/10.4102/jamba.v9i1.391

Lewis, J., & Kelman, I. (2012). The good, the bad and the ugly: Disaster risk reduction (DRR) versus disaster risk creation (DRC). *PLOS Currents*, *4*. https://doi:10.1371/4f8d4eaec6af8

Lund, C. (2006). Twilight institutions: Public authority and local politics in Africa. *Development and Change*, *37*(4), 685–705. https://doi.org/10.1111/j.1467-7660.2006.00497.x

Marchezini, V., & Wisner, B. (2017). Challenges for vulnerability reduction in Brazil: Insights from the PAR framework. In V. Marchezini, B. Wisner, L. Londe, & S. Saito (Eds.), *Reduction of vulnerability to disasters: From knowledge to action* (pp. 57–96). Rima Editora.

Müller, M., Penny, G., Niles, M., Ricciardi, V., Chiarelli, D., Davis, K., Dell'Angelo, J., D'Odorico, P., Rosa, L., Rulli, M., & Mueller, N. (2021). Impact of transnational land acquisitions on local food security and dietary diversity. *Proceedings of the National Academy of Sciences*, *118*(4), e2020535118. https://doi.org/10.1073/pnas.2020535118

Murphy, A., & Contreras, I. (2022, May 12). The global 2000. *Forbes*. https://www.forbes.com/lists/global2000/?sh=6dda3c75ac04

Ndikumana, L., & Boyce, J. (2011). *Africa's odious debts: How foreign loans and capital flight bled a continent*. Zed Books.

Newmiller, J., Walker, W., Fleenor, W. E., & Pinter, N. (2020). Case study: Reconstructing the 2015 Dulcepamba river flood disaster. *Environmental & Engineering Geoscience*, *26*(4), 393–404. https://doi.org/10.2113/EEG-2337

O'Keefe, P., Westgate, K., & Wisner, B. (1976). Taking the naturalness out of natural disasters. *Nature, 260*(5552), 566–567.

Oliver-Smith, A. (1994). Peru's five hundred year earthquake: Vulnerability in historical context. In A. Varley (Ed.), *Disasters, development and environment* (pp. 31–48). John Wiley & Sons.

Oliver-Smith, O., Alcántara-Ayala, I., Burton, I., & Lavell, A. (2016). *Forensic investigations of disasters (FORIN): A conceptual framework and guide to research. IRDR FORIN publication no. 2.* Integrated Research on Disaster Risk.

Organisation for Economic Co-operation and Development. (2014). *OECD foreign bribery report: An analysis of the crime of bribery of foreign public officials.* OECD Publishing.

Peng, C., Regmi, A., Qiang, Z., Yu, L., Xiaoqing, C., & Deqiang, C. (2017). Natural hazards and disaster risk in one belt one road corridors. In M. Mikos, B. Tiwari, Y. Yin, & K. Sassa (Eds.), *Advancing culture of living with landslides* (pp. 1155–1164). Springer.

Piciullo, L., Storrøsten, E., Liu, Z., Nadim, F., & Lacasse, S. (2022). A new look at the statistics of tailings dam failures. *Engineering Geology, 303*, 106657. https://doi.org/10.1016/j.enggeo.2022.106657

Quijano, A. (2000). Colonialidad del poder y clasificacion social. *Journal of World-Systems Research, 1*(2), 342–386.

Richter, B., Postel, S., Revenga, C., Scudder, T., Lehner, B., Churchill, A., & Chow, M. (2010). Lost in development's shadow: The downstream human consequences of dams. *Water Alternatives, 3*(2), 14–42.

Samuel, K., Aronsson-Storrier, M., & Bookmiller, K. (2019). *The Cambridge handbook of disaster risk reduction and international law.* Cambridge University Press.

Sanderson, D., Patel, S., Loosemore, M., Sharma, A., Gleason, K., & Patel, R. (2022, October). Corruption and disasters in the built environment: A literature review. *Disasters, 46*(4), 928–945. https://doi.org/10.1111/disa.12500

Scudder, T., Manley, R., Coley, R., Davis, R., Green, J., Howard, G., Lawry, S., Martz, D., Rogers, P., Taylor, A., Turner, S., White, G., & Wright, E. (1993). *IUCN review of the Southern Okavango integrated water development project.* Samara Publishing Company for IUCN. https://portals.iucn.org/library/sites/library/files/documents/WTL-015.pdf

Sharon, S. (2022). Supreme court of India and environmental justice: A historical analysis. *Journal of Polity and Society, 14*(2), 137–148.

Siciliano, G. (2014). Rural-urban migration and domestic land grabbing in China. *Population, Space, Place, 20*, 333–351. https://doi.org/10.1002/psp.1830

Siddiqi, A. (2022). Disaster studies and its discontents: The postcolonial state in hazard and risk creation. In G. Bankoff & D. Hilhorst (Eds.), *Why vulnerability still matters: The politics of disaster risk creation* (pp. 91–106). Routledge.

Staub, M. (2019, March 19). Devastating floods in the aftermath of cyclone Idai. *ReliefWeb.* https://reliefweb.int/report/mozambique/devastating-flooding-aftermath-cyclone-idai

Transparency International. (2016). *The European corruption observatory: Investigating corruption.* Transparency International. https://transparency.eu/wp-content/uploads/2016/10/TIE-003-16_web.pdf

United Nations Environment Programme. (2023, January 24). *New independent global tailings management institute announced to drive mining industry safety standard.* United Nations Environment Programme. https://www.unep.org/news-and-stories/press-release/new-independent-global-tailings-management-institute-announced-drive

United Nations Human Rights Council. (2011). *Report of the special rapporteur on adequate housing as a component of the right to an adequate standard of living, and on the right to non-discrimination in this context.* United Nations Human Rights Council.

United Nations Human Rights Council. (2015, December 8). *Brazil mine disaster: UN human rights expert calls for urgent access to safe drinking water.* United Nations Human Rights Council. https://www.ohchr.org/en/press-releases/2015/12/brazil-mine-disaster-un-human-rights-expert-calls-urgent-access-safe

Venter, O., Sanderson, E., Magrach, A., Allan, J., Beher, J., Jones, K., Possingham, H., Laurance, W., Wood, P., Fekete, B., Levy, M., & Watson, J. (2016). Sixteen years of change in the global terrestrial human footprint and implications for biodiversity conservation. *Nature Communications, 7,* 1–11. https://doi.org/10.1038/ncomms12558

Wagret, P. (1968). *Polderlands.* Methuen and Co.

Waldmüller, J., & Robledo, M. (2021). ¿Descolonizar internamente? Rastreando los patrones militares y colonizadores en Ecuador a través de la gestión de riesgos. *Revista de Estudios Latinoamericanos sobre Reducción del Riesgo de Desastres, 5*(2), 34–47. https://doi.org/10.55467/reder.v5i2.69

Wisner, B. (1993). Disaster vulnerability: Scale, power and daily life. *GeoJournal, 30,* 127–140. https://doi.org/10.1007/BF00808129

Wisner, B. (2003). Changes in capitalism and global shifts in the distribution of hazard and vulnerability. In M. Pelling (Ed.), *Natural disasters and development in a globalizing world* (pp. 43–56). Routledge.

Wisner, B. (2020). Five years beyond Sendai: Can we get beyond frameworks? *International Journal of Disaster Risk Science, 11,* 239–249. https://doi.org/10.1007/s13753-020-00263-0

Wisner, B., Blaikie, P., Cannon, T., & Davis, I. (2004). *At risk: Natural hazards, people's vulnerability and disasters* (2nd ed.). Routledge.

Wisner, B., & Gaillard, JC. (2009). An introduction to neglected disasters. *Jàmbá: Journal of Disaster Risk Studies, 2*(3), 151–158.

Wisner, B., Gaillard, JC., & Kelman, I. (2012). Framing disaster: Theories and stories seeking to understand hazards, vulnerability and risk. In B. Wisner, JC. Gaillard, & I. Kelman (Eds.), *The Routledge handbook of hazards and disaster risk reduction* (pp. 18–34). Routledge.

Wisner, B., & Lavell, A. (2017). *The next paradigm shift: From 'disaster risk reduction' to 'resisting disaster risk creation'* (Paper presentation). Dealing with Disasters - University of Durham, United Kingdom.

World Bank. (2022, April 19). Mozambique's hidden debt: Turning crisis into opportunity. *World Bank Blogs.* https://blogs.worldbank.org/africacan/mozambiques-hidden-debts-turning-crisis-opportunity-reform

Yang, B., & He, J. (2021). Global land grabbing: A critical review of case studies across the world. *Land, 10*(3), 324. https://doi.org/10.3390/land10030324

Chapter Ten
Understanding Risk and Living Life

The Progression of Safety

The final chapter of our book returns to the PAR framework and reverses it. The progression of vulnerability becomes "the progression of safety" (POS). In order to enjoy safe conditions of life, a person must have the capacity and ability to assure satisfaction of basic human needs (Wisner, 1988)[1] including site, situation, and stability of shelter despite exposure to hazards, and to thrive on the basis of reliable and sustainable livelihoods.

A new set of dynamic pressures will be discovered by reversing the progression of vulnerability. These new dynamic pressures lead back to and reveal a revised set of root causes that overturn historical oppression, marginalization, expropriation, and exploitation; heal the societal wounds and fragility created by prior wars and crises; and seek ways of favoring equitable distributions of power, resources, and wealth. The reader may find such "overturning", "healing", and "favoring" overly ambitious and ringing too much of the prophet Isiah set to music by Handel. We have no such utopian visions; rather, the new cascade of root causes and dynamic pressures signals a direction for modified policy and practice if and when there is sufficient political demand for the construction of safety. The reversed framework suggests a direction for change, not the speed of change, which will have to be negotiated among everyone. Later in this chapter, we

DOI: 10.4324/9781003292814-10

discuss that negotiation process in the light of the 2019 Manifesto and 2021 Accord signed by hundreds of disaster researchers and activists (https://www.radixonline.org/manifesto-accord). The Manifesto and Accord insist that ordinary people be central to the production of safety and control the process.

We call this chapter "Understanding Risk and Living Life". What to make of the second phrase in the title? Much writing about disaster risk focuses on "living with risk" as in the expression "living with landslide risk" or "living with flood risk", not on the living of life beyond risk. People live their lives in systems of power: economic, social, administrative, political, and coercive. We earn, grow, manufacture, buy, and sell; we visit a sick friend, seek a neighbor's permission to cut a tree, and license our pet dog (where required); we vote (where we are permitted to), pay taxes, and obey laws; and we challenge bullies and avoid parts of town where coercive violence is frequent. Only the most skillful and fortunate hermit can avoid living in this way, immersed in and fully dependent on systems of power. Many of these systems of power are invisible to us. We are seldom consciously aware of our participation in these systems and our dependency on them. If reminded, most people would be as surprised as the gentleman in Molière's (1670) play who was amazed he had been speaking prose for many years.

We should be all the more amazed that ordinary lives reproduce and reinforce those systems of power. It takes a conscious change in how daily life is lived in order to reverse the progression of vulnerability. Note what changes Figure 10.1, the release model, suggest: redistributing all those kinds of power; redistributing wealth and access to resources; and overturning historical race, gender, caste, and other forms of oppression and colonized ways of thinking, as well as the healing of societal wounds and fragility. It remains for this chapter to explore if and how conscious, incremental changes in living our ordinary lives can overturn those systems of power. Can such efforts yield interdependency within groups of people, joy, and fully meaningful life?

Local Actions to Achieve Safe Conditions

As hard as it is to reverse the root causes listed at the extreme left of Figure 10.1, conscious, incremental change might begin with addressing unsafe conditions and dynamic pressures through local efforts to mitigate hazards. Safe conditions are divided into those to do with natural resources, those involving social organization, and others requiring public action. Thinking about the items mentioned under the heading "natural resources", groups of people could unilaterally broaden access to land and water despite absentee ownership. People are already doing that as land occupations in Latin America have shown (Harris & Ingizza, 2023). Likewise, local initiatives to protect biodiversity are

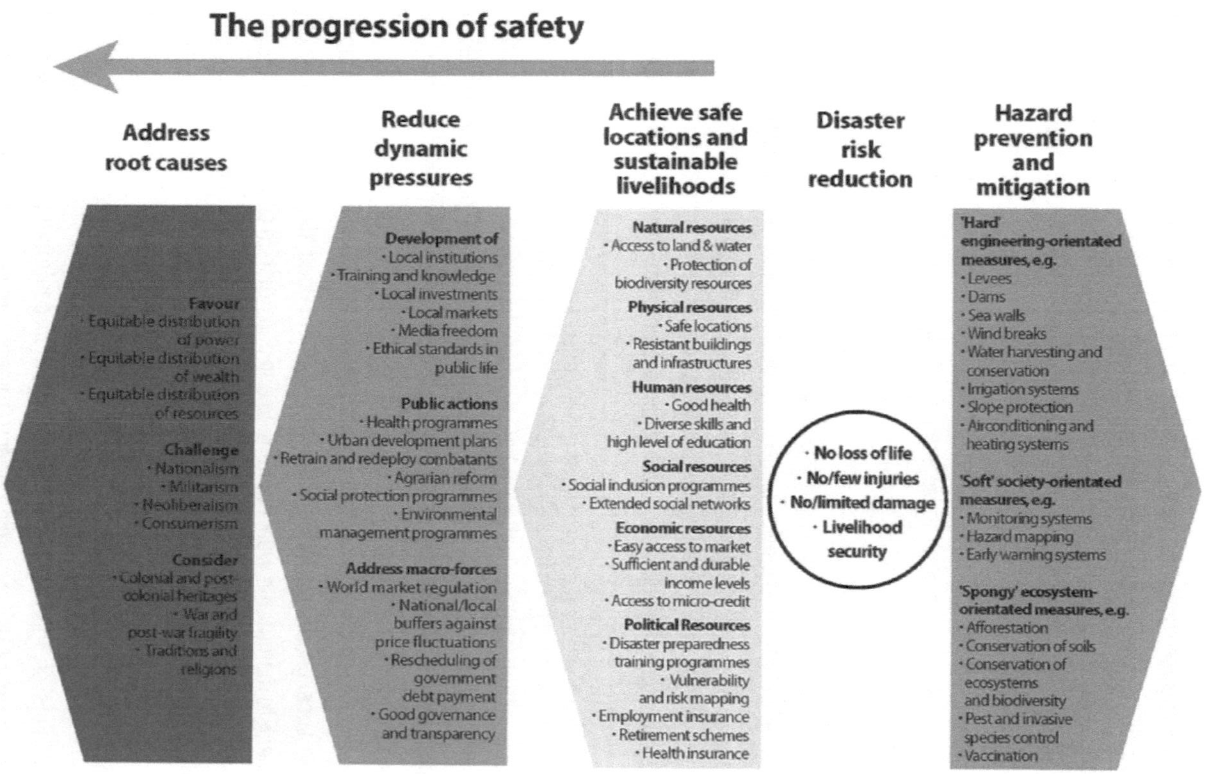

FIGURE 10.1 Progression of safety

Source: Wisner et al. (2012)

The Progression of Safety

possible and currently underway (Corlett, 2020; Peter et al., 2021). As cities and regions implement renewable energy plans and demand reduction, people will gain income opportunities, safer jobs, more secure livelihoods, and cleaner environments.

The safe conditions specified under the heading "socio-economic" are also amenable to local group action. Social inclusion can be increased by dialogue among groups historically divided by race, religion, language, immigration status, and other characteristics. Local economic production, circulation, and consumption can result from partially de-linking from non-local sources of food and other goods for which there are local or regional substitutes for international imports. Such changes can also potentially enhance diversity of local incomes and livelihoods (Bello, 2004). Even some of those actions mentioned under the heading "public" need not wait for top-down government – that is, officially public – action. Risk mapping and preparedness require only awareness and willingness of a group of people, including some with certain levels of literacy and numeracy. The oldest members of a community might have sufficient recall of hazards to provide some sense of places and livelihoods at risk. Such knowledge must be considered in tandem with externally collected data, maps, archives, and other forms of knowledge, many available with a tablet and internet connection. Insurance, too, is possible to a degree if people build upon shared experience of traditional mutual aid and are able to extend their network to others in regions prone to a different set of hazards or that live with the same hazards but on a different seasonal calendar.

The limit of local action can be reached when the POS framework calls for a sophisticated item, such as a vaccine or engineered dam. Nevertheless, people themselves can plant windbreaks, reafforest small areas, and support vegetation that absorbs storm surges. Traditional forms of irrigation, such as flood-retreat in which a succession of crops is grown after seasonal river flooding, exist to inspire the imagination and actions of people today. Local groups can inform themselves about good planning and construction and then seek to monitor the work of planning and construction companies. They can endeavor to educate others and push back against rumors and false information, such as interests producing fear of vaccination.

Local Actions to Counter Dynamic Pressures

Less seems feasible on the basis of local action when confronted with the dynamic pressures listed in Figure 10.1. So-called "macro-forces" require international debt management and honest, transparent government. Vigorous participation in local, regional, national, and international politics is a critical requirement for local groups, assuming the existence of a democratic system and the time and resources to be an active participant. Other actions countering dynamic pressures are possible locally, such as filling organizational gaps by creating new institutions, establishing local markets, educating, and training, as well as some limited public health activities, such as digging latrines and improving drainage. Another example is basic health care based on the model of the "barefoot doctor" (village medical helper) once championed by the World Health Organization. Limits again appear when it comes to planning, building, maintaining, and regulating infrastructure and the built environment in places where people do not have these options, where planning and construction are unregulated, and where the wider processes seek to continue constructing disaster risk, as discussed in Chapter 9.

A Vingtième in the Blink of an Eye: 20 Years Pursuit of Safety

A now-classic source and discussion of PAR ends with a long chapter entitled "Towards a safer environment" (Wisner et al., 2004, pp. 330–373), where details of a hoped-for "reversal" are laid out:

> We are optimistic about the possibility of improvement. The "Pressure" model can be reserved to provide security instead of risk. Vulnerable people's access to resources can be improved, and changes in power relations can be made, vulnerability can be decreased, and if aid is properly conceived and implemented even the most vulnerable survivors [of a disaster] can recover in such a way that future vulnerability is reduced.
>
> (Wisner et al., 2004, pp. 342–343)

Were Wisner et al. (2004) overly optimistic? Looking for the root causes and dynamic pressures that lead to disasters is necessary if one wants to know what is necessary to reverse PAR and to construct safety. Thus, concerning whether the optimism expressed at the end of the second edition of *At Risk* was excessive, the verdict "not proven" may have to apply. We have to dig deeper to make that assessment beginning by asking what has changed in the 20 years since Wisner et al. (2004) wrote those words. That long final chapter of *At Risk*'s second edition reviews a series of international conferences on disaster risk reduction and finishes with an exhaustive discussion of seven risk reduction objectives culled from many sources. These seven objectives seemed feasible at the time, but they need to be reassessed a couple of decades later.

The COVID-19 pandemic was a reminder that human, animal, and ecosystem health are tightly interwoven (Lawler et al., 2021; Singer, 2023). Human encroachment into formerly unused regions has accelerated. As Watson et al. (2018) describe the results of their global mapping project, "a century ago, only 15% of Earth's surface was used to grow crops and raise livestock. Today, more than 77% of land (excluding Antarctica) and 87% of the ocean has been modified by the direct effects of human activity". Fossil fuel use had continued to grow despite conferences and pledges. The intensity of storms is increasing. The United Nations is not optimistic. Its *Global Assessment Report on Disaster Risk Reduction* for 2020 entitled "Our World at Risk" (United Nations Office for Disaster Risk Reduction, 2022) warns, "COVID-19 and climate change are rapidly making it clear that, in today's crowded and interconnected world, disaster impacts increasingly cascade across geographies and sectors. Despite progress, risk creation is outstripping risk reduction".

Another disturbing trend is continuing contempt for science and denial of scientific evidence. Witness politically charged vaccination refusal during the COVID-19 pandemic (Chapter 4) and denial of human-caused climate change by populists, including one US president (Marquardt & Lederer, 2022). Violent conflict compounds some of these trends by making it difficult or impossible for governments or international organizations to operate in contested territories. Conflict has directly exposed a growing number of people to food insecurity and indirectly exacerbated that situation through rising costs of fuel and fertilizer (Mia, 2022). Wealth has continued to be concentrated. The poorest half of the world's adults (2.6 billion individuals) own only 2% of the world's wealth, while "Since 1995, the share of global wealth possessed by billionaires has risen from 1% to over 3%" (World Inequality Report, 2022, p. 15) and, over the same time period, "The share of wealth owned by the global top 0.1% rose from 7% to 11%" (World Inequality Report, 2022, p. 3).

Finally, life itself and meaning have been colonized by corporations seeking legal ownership of DNA (Wynberg, 2023) and by generative artificial intelligence (AI) that threatens to flood a digitally interconnected world with simulacra of the expository history, biography, poetry, drama, and other works formerly authored by human beings. Should this be seen as an assault on culture as the mirror in which living human beings look in order to understand themselves or as a 21st-century addition to human culture with huge value – or both? Whatever the answer(s), in practical terms, safety and security concerns abound about legal ownership of genetics and AI. These include causing harm by mistakes in biomedical and pharmaceutical procedures, unregulated surveillance of legalities and technologies, automatic indiscriminate killing by drones and missiles, and AI-driven autonomous weapons which can be highly targeted or left without much instruction.

Autocritique and Revision

With the weight of two decades' challenges on our shoulders, we now return to the seven proposals that concluded the second edition of *At Risk* (Wisner et al., 2004). In this light, they could read as naïve and cheerfully optimistic.

1. Understand and communicate the nature of hazards, vulnerabilities, and capacities (pp. 330–333)

   It will be increasingly difficult to tie specific groups of people to specific or small numbers of vulnerabilities. Information will be missing. Misinformation and disinformation will abound. People will be subject to meteorological and geopolitical storms of increasing violence. It will be less clear to whom those who "understand" are supposed to "communicate" in a world populated by an increasing number of states with inadequate governance and governments.
2. Conduct risk assessment by analyzing hazards, vulnerabilities, and capacities (pp. 333–342)

   Such assessments are challenging and often not useful in conflict zones. They are difficult in situations where people have grown weary of researchers, "experts" (often self-appointed), and remote government officials telling them what they should do. The obstacles to so-called "community participation" were addressed in Chapter 3. Nevertheless, in some places, it remains feasible for at-risk people themselves to conduct such assessments and to use them to lobby for risk reduction and safety.
3. Reduce risks by addressing root causes, dynamic pressures, and unsafe conditions (pp. 342–348)

   The key to any attempt to reverse PAR and to promote the construction of safety and meaningful living is the struggle with root causes of vulnerability. It was a shorter list in the first edition of *At Risk* in 1994 than in the second edition of 2004 since root causes had proliferated and the ability to identify them had improved. If these trends continue, nothing short of changing corporate and state power-hungry monoliths into simpler, more directly democratic, transparent, accountable, responsible, multi-scalar approaches would seem to be a feasible solution.
4. Build risk reduction into sustainable development (pp. 348–350)

   The phrase "sustainable development" has been hijacked and recast as "sustainable growth", an obvious oxymoron. Nevertheless, with tweakings and multiple tech fixes along the way, techno-optimists still believe that the existing market-based, profit-above-all-else, focus-on-the-short-term system can avoid shipwreck on the rocks and shoals of environmental destruction, including human-caused climate change, war, and growing inequality. We are profoundly skeptical that the current system can achieve what is needed.
5. Reduce risks by improving livelihood opportunities (pp. 351–353)

   Small-scale, locally-based, and controlled efforts have been successful. They tend to be fragile, often dependent on outside funding and advocacy, can take place in the shadows of failed states, and are under threat of kleptocrats demanding rent or other favors in the absence of which sponsors can be expelled from the country.

6. Build risk reduction into disaster recovery (pp. 353–367)

   Again, this is feasible and has been successful. The question remains whether "recovery" turns out to be a return to the status quo with all of its associated hazards, vulnerabilities, and blocks to resistance and capacity. That is, does it just "return to normal" – the normal of vulnerability leading to disasters?

7. Build a safety culture (pp. 367–374)

   The idea of building a safety culture is frequently articulated in a Western sense. It is based on concepts and values that do not always have counterparts in other cultures. Therefore, the construction of a culture of safety as often presented is doomed as a universalizing project and can only partially and with difficulty be instantiated in non-Western cultures, where there are Western-trained or Westernized leaders (government, corporate, nonprofit, or academic). To understand the difficulty facing the transfer of the idea of safety culture, consider a common definition (Health and Safety Commission, 1993, p. 23):

> The product of individual and group values, attitudes, competencies and patterns of behavior that determine the commitment to, and the style and proficiency of, an organization's health and safety programmes. Organizations with a positive safety culture are characterized by communications founded on mutual trust, by shared perceptions of the importance of safety, and by confidence in the efficacy of preventive measures.

This definition assumes that individual values can be distinguished from group values. It juxtaposes "attitudes", "competencies", and "patterns of behavior" that would be difficult or impossible to distinguish in many cross-cultural contexts. Likewise, "style" and "proficiency" can be overly individualistic for easy transfer. "Commitment" is a concept with one anchor in 19th-century Western notions of duty and deontological ethics. Many of these words, values, and intentions exist in numerous cultures; many do not. The issue is not throwing out all these ideas but rather ensuring that they make sense, translate, transfer, and can be implemented in many contexts, even if (especially if) requiring a major re-framing, re-working, and/or re-interpretation.

Reception and Use of PAR and Possible Role for a Progression of Safety (POS) Framework

The PAR frameworks from both editions of *At Risk* have been much used and cited, although the "pressure" form is better known and more frequently cited and applied than its "release" counterpart. The reason for this disparity is that more academics than policy makers have been attracted to this body of work on disaster. Academics are more interested in unpacking and understanding the causes of disaster risk. That is the strength of the progression of vulnerability. By contrast, the progression of safety is more useful in the work of reform and restructuring of institutions and the processes of governance. This is not the sort of critical activity of the kind that academics relish. Rather, it involves the hard slog of incremental change, usually promoted by advocates within institutions. It remains a challenge to those working with PAR to find more effective and locally relevant ways of introducing the progression of safety into the protocols, routines, and practices of civil servants and elected and appointed officials. The latter are found everywhere from local councils that govern hamlets and small towns to regional and national governments. In addition, there are decision-makers who populate international organizations and their counterparts in non-governmental organizations and the private sector.

Both "pressure" and "release" (progression of safety) versions of PAR are flexible frameworks, meant to encourage questions. As such, they are easily revised and updated. In the case of the progression of safety, dynamic pressures might be added and the root causes may change. Examples include human-caused climate change and the unintended risks produced by investments made to reduce greenhouse emissions and increase their uptake. Likewise, "patriarchy, racism, and colonialism" might be added to root causes, among other forms of discrimination and "otherism". "Social protection" could be added to the array of institutional anchors of safety under dynamic pressures as well as access to safe housing and access to credit.

Our journey toward safety will have to be a joint effort. Addressing the root causes of vulnerability indeed entails that those with power focus on shifting power structures and granting access to resources to those most at risk. Yet the progression of safety also requires building upon, maximizing, and sharing people's capacities, capabilities, and abilities (Chapter 3). As such, efforts to make people safe must be localized yet integrated with initiatives at different and broader (regional, national, and international) scales. Making sure that actions geared toward addressing the root causes of vulnerability and efforts focused on people's capacities, capabilities, and abilities come together and work toward the same goal is one of the greatest challenges of disaster risk reduction (Wisner et al., 2012; Gaillard & Mercer, 2013).

The current impetus toward localization for disaster risk reduction initiatives is a welcome development in this direction. It may easily be co-opted by those holding power in the development "industry", hence, reinforcing asymmetrical power relations and doing little to foster culturally grounded approaches to reducing disaster risk. Localized disaster risk reduction requires more than just working at the local level with local people and local stakeholders. It entails understanding harm, hardship, and suffering through local understandings of the world and within the context of people's daily life.

No matter how often and thoroughly one might revise and update the progression of safety, the task of understanding risk must at some point look beyond risk and concern itself with understanding life as it is lived in the 21st century. There are at least two reasons why this broader perspective is required. First, in its classic forms (Blaikie et al., 1994; Wisner et al., 2004, 2012), PAR has limited place for individual experience of risk or experience of safety. These original sources of PAR and POS are only barely able to accommodate the ways different cultures understand individual experience as it manifests as anxiety, fear, terror, resignation, watchfulness, and comfort. These emotional states and subjectivities will also likely vary by gender, race, sexual orientation, and other identity-based standpoints. Such variation must also be accommodated, and it is not mentioned in the classic discussions of PAR.

The second reason why this chapter has to go beyond risk and safety and their abstract representations in the PAR framework has to do with human aspirations. The COVID-19 pandemic caused a temporary reduction in geographical mobility and, for some, a longer-term shrinking of social life and even imagination, as well as the circle of one's curiosity and empathy. Still for many people, to be "safe" is not their ultimate desire but rather to thrive, to be happy, and to care for and make others happy. As people emerged from pandemic-induced relative isolation, they sought to reestablish sociality and creative activity. Many are now fortunate enough to be able to assume their physical safety as a background condition. Some living in violent neighborhoods may have to think twice about venturing out to the shops or may be worried about their child's journey to and from school. Those living in war zones and regions of social instability, such as Haiti, Gaza, and the Democratic Republic of the Congo, must have daily safety on their minds. The fortunate can assume that these are exceptional situations. How long such fear and worry will remain exceptions is a relevant question. Ultimately, we all are at risk and have ways of dealing with that risk which might not always be "safety".

Reflecting on experience during the COVID-19 pandemic, Butler (2022) asked, "What kind of world is this? . . . How is one to live in such a world?" (p. 29). Her inquiry led her to question what kind of world is "inhabitable" and for whom. Butler (2022) believes an ethical and political project is possible that would create a common world in which life is livable:

> a politics that is committed to a world in which we can all breathe without fear of contagion, fear of pollution, or fear of the police chokehold, where our breath is intermingled with the world's breath, syncopated and free, becomes what is shared – our commons.
>
> (p. 109)

This "ethical and political project" would define the conditions under which such a commons would exist and for whom. Butler asserts that the pandemic taught lessons about when and how lives are grievable – a theme Butler had been treating for some time (Butler, 2009) – and also gave rise to networks of care in the form of mutual aid societies and nonnuclear pods during 2020–2021. She argues that these emergent responses to the pandemic should not be considered "as a private and sequestered activity but as a form of power with the potential to change global practices and institutions and to transform the world" (Butler, 2022, p. 106).

A more somber view is expressed by Latour (2018), who describes politics in what he names the "New Climate Regime". He means "climate in the broad sense of the relations between human beings and the material conditions of their lives" (Latour, 2018, p. 1). He believes that widespread deregulation, rampant and growing inequality, and denial of (human-caused) climate change have resulted in the loss of any common, shared world. This loss has two dimensions, according to him. First, human beings are increasingly separated from one another, not communicating and less curious about one another's values and situations. Second, human beings grow ever more destructive of other beings that constitute the material world with less awareness of the impact of their consumption-destruction. The way to re-create a common world is to rethink the relationship between humans and the other living and non-living beings that make up the world, which he names Gaia, using James Lovelock's term (Latour, 2017, pp. 288–289). Latour is more pessimistic than Butler, and his solution to the "madness" he perceives around him, the unanchored world he describes in *After Lockdown* (Latour, 2021), is intellectually and culturally more demanding. This is because his solution would involve extending the "conference of the parties" – such as the familiar "COPs" of more than two dozen international climate change negotiation processes – to non-human beings, even though it would be a demanding process (Latour, 2017).

The reversal of PAR must be generative of questions about what is required to make a world inhabitable and provide conditions in which a life is livable in the face of not only hazards but also many other forms of precarity, insufficiency, and failure. For instance, Latour's (2017) focus on "living and non-living beings" should be expanded to include ecosystems and biomes. He and others suggest that denial of (human-caused) climate change is one symptom of the denial of evidence and respectful exchange. Delusion and error breed where people separate themselves from others and refuse to listen. Yet the explosion of social media raises another possibility. We are both increasingly separated from each other and simultaneously too "connected" in an inhuman manner. We are excessively "wired" and receive too much information and imagery that demand our attention. This comes too often, too rapidly, and with too many expectations of instant responses. Automated trolls powered by AI have the potential to escalate these deluding, scattering, and fragmenting "messages" by orders of magnitude.

Living on the Earth in the Circle of Abilities

Human beings have capacities or abilities that allow them to understand themselves and each other to some degree and to engage in various ways with their more-than-human surroundings (living and non-living) with both intended and unintended consequences. As living beings, humans are part of the web life and the thin layer of planetary life known as the biosphere. Teilhard de Chardin (2008) believes that humans have also created a concentric layer of thought, emotion, and belief which he calls the noosphere that interpenetrates the biosphere. "*Noos*" comes from the classical Greek word for mind. Without reifying humans' ability to learn and to transmit accumulated learning across generations as *noos*, we will use the less lofty word "culture". Culture is what provides humans with capacities to use social, economic, political, human, physical, and natural resources to modulate their engagements with the more-than-human in order to construct safety. In Chapter 9, we used the triangle of vulnerability to show how the misuse of these resources may lead to marginalization and vulnerability and the construction of disaster risk.

Now in Chapter 10, we see what is required for people's capacity to cooperate with each other and to create an environment that can drive both the progression of safety (POS) and the progression of meaningful livelihood and living (PLL). A recluse living in a Cold War bunker may be "safe" to some extent, while others might query (legitimately or otherwise) how much they are living a meaningful life – with the key question being "meaningful for whom?" In this manner, our concluding chapter pursues the construction of safety and meaning (CSM) as the opposite of the construction of disaster risk (CDR). Psychologically, anxiety is the emotion associated with POS on its own, looking over one's shoulder with fear of disaster. That can be life-saving in some circumstances. Life should presumably be more than the avoidance of risk, and the emotions associated with PLL and CSM are joy and confidence. The resources and some of the tools for groups of people to begin CSM and to enjoy PLL are shown in Figure 10.2.

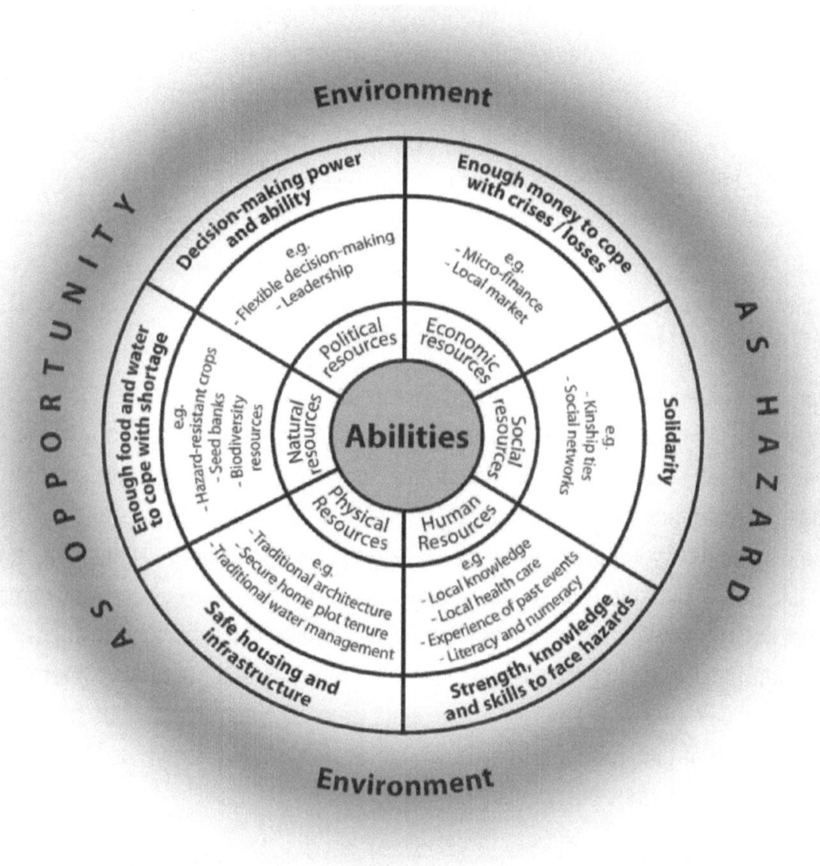

FIGURE 10.2 The circle of abilities

Living Life Beyond Disaster Risk

We need to move beyond the concept of disaster risk without losing sight of it, in order to consider everyday risks, and then, ultimately, to navigate entirely beyond risk by creating the systems and structures which people require and deserve to give ourselves options for life and living (livelihoods, yes, but also opportunities for leisure and self-expression). The phrase from the last chapter of *At Risk*, "Create the systems and structures that people require and deserve", puts human agency in the spotlight. We inhabit a planet with a molten core, shifting tectonic plates, a turbulent atmosphere, constantly churning oceans, and encountering phenomena from outer space. We draw our livelihoods directly and indirectly from these dynamisms. We label those that sustain us as "resources" during our engagements with more-than-human life and our non-living environments in the ways discussed in Chapter 4. These dynamisms can harm us. The harms, we label "hazards". We have only partial understanding of these biotic and abiotic dynamisms. The power of these processes is difficult to control and perhaps should not be entirely controlled. We and all that we have created for our lives can easily be buried by landslides, incinerated by wildfires, shaken to destruction by earthquakes, desiccated by droughts, washed away by floods, ingested by wildlife, or sicken and die from disease.

"Understanding Risk" So Far

Chapter 1 argued that to understand disasters is challenging because their multiple causes accumulate slowly and involve the use and abuse of power in many forms: economic, political, socio-cultural, administrative, and coercive. All these are societal processes that common sense accepts as "normal", as are the decisions and actions of those who hold power regarding how we behave and how we relate to each other and to non-human life. Our abiotic surroundings and non-human life are as diverse and complex. It is not easy to anticipate whether the lithosphere, hydrosphere, biosphere, atmosphere, cryosphere, and outer space may manifest as harmful. Society and our planet change from day to day, season to season, year to year, spanning vast time periods covering the life and death of generations. Hazards change as do our vulnerability to them and our ability to modify hazards or to reduce our exposure to them.

Chapter 2 reviewed several frameworks for understanding disaster risk. Some of them discuss the role of power and marginalization and others do not. One of the frameworks that engages with power and its abuse is "the progression of vulnerability" or "pressure and release" framework (PAR). PAR distinguishes among processes that are more remote and less remote in time and space and then traces the cascading consequences of remote processes as they are focused by local, situational pressures and eventually influence immediate safety conditions. Among the root causes are several such as discrimination and exploitation that involve the use and misuse of power and resources. The survey further revealed several kinds of applications of PAR. These included sector- and hazard-based applications (e.g., health- or earthquake-related), applications by international and national organizations, and decentralized use by local governments and nongovernmental organizations. Many featured local people's participation.

In Chapter 3, we explored people's abilities to tackle elements of PAR. Notions of ability, capability, capacity, coping, and recovery among others were examined and critiqued, especially from various cultural and institutional perspectives. We discussed views of people's ability to resist macro-forces and oppressive structures that constitute the root causes of vulnerability in the PAR framework. The traditions of objectivism and structuralism suggest that people's behaviors are constrained by structural forces beyond their reach, such as the root causes of vulnerability. This leads to consideration of unequal power relations as a factor preventing people from exerting their ability to deal with hazards and disasters. By contrast, subjectivist and constructivist traditions, including existentialism, have argued that people are always able to express their free will. The social fabric results from individual decisions. In this way, people's abilities to deal with hazards and disasters would be independent from the overall social fabric and could be expressed as acts of self-determination.

Caution is required. Abilities are not superpowers as much as one would like to recruit fictional characters or deities to serve as disaster risk managers. In the world in which we actually live, abilities/capacities are performed in the course of actual lives and life projects.

Diverse non-human life was taken up in Chapter 4, a journey that took us to both ends of the scale continuum, from microbe to elephant, as it reflected on our relations with the rest of the biosphere. Be it disease or digestion, such duality (not binary) plays out in our daily lives at the microbiological scale, as well as in our relations with plants and animals at the macrobiological scale. Some of these plants are toxic to humans, and some nourish domesticated animals. Others are labeled "invasive" because they proliferate where humans want to grow food. Animals compete for our plant food: birds and insects eat it in the fields or storage areas, and others trample crops or make their homes in it. As living beings, humans have an outsized impact on the rest of the biosphere that has meant the extinction of species, a process continuing at rapid rates.

In Chapter 5, human relations with their abiotic surroundings were shown to be complex. Many of the same processes can be experienced by humans as potentially harmful (hazards) or potentially useful or beneficial (resources) – and sometimes simultaneously. Chapter 5 involved a mental journey through the atmosphere, touching on weather and climate impacts on humans and, in the case of climate, how human activity has been causing the Earth's atmosphere and oceans to heat. The hydrosphere-atmosphere links are as critical conditions for life as links among chemical elements, such as nitrogen, phosphorus, and carbon – elements that originated in the supernova deaths of stars billions of years ago. Human beings have adapted to the seasonality of weather and depend on the fluctuation of heat, humidity, rainfall, and wind, even while any can be deadly. We saw that natural climate change has taken place in the past as a function of the changes in the Earth's orbit around the Sun. Human-caused climate change is a byproduct of our consumption activities, especially our use of fossil fuels as a source of energy.

In Chapter 6, we saw that even the Earth's molten core and mantle are essential to human and non-human life. The spinning molten core creates a magnetic shield around the earth that protects all life from damaging cosmic radiation. Where the magma comes to the surface, volcanoes may devastate our infrastructure, yet volcanic ash provides fertile soil that attracts farmers. Geothermal energy may be obtained from the heat below. Our planet's lithosphere floats as stony plates on the molten interior of the Earth. These plates move and collide; they ride over and under or slip past one another. Such tectonic movement gives rise to earthquakes but sometimes create aquifers from which we have learned to extract freshwater.

Chapter 7 provided additional insight into the dynamism of earth and ocean systems. Landscapes are formed – and become homes and useful to humans, as well as hazardous when slopes collapse and slip. Earthquakes and massive shifting of slopes on the seabed among other phenomena cause tsunamis that may have the potential to scour low-lying lands, sweeping away vegetation and infrastructure.

Chapter 8 took us into the rest of our home solar system and galaxy, where cis-lunar and more distant influences on human life exist and can be experienced by humans as resources or hazards. Solar radiation drives photosynthesis, thereby empowering all life on Earth, but could wreak havoc on Earth's life with a solar flare and other space weather. Our moon helps create ocean tides and is, therefore, responsible for creating ecozones, where a rich diversity of life thrive. Near-Earth objects that come to and go from our solar system, such as asteroids and comets, could collide with our home planet, causing great damage. More distant from the near and now, events in the galaxy billions of years ago seeded our world with chemical elements that make up our bodies and the flesh of all plants and animals. Also produced by nuclear fusion in the hearts of those dying stars were the rare elements that humans have mined for thousands of years and continue to find new uses for.

Chapter 9 opened our eyes to the scandal and tragedy of disaster risk construction. International and national attempts to reduce disaster risk are limited by their failure to address the misuse of power and other root causes of vulnerability. Risk is intentionally created by elites acting to control resources. These elites that control resources, choices, and opportunities typically abuse and misuse their power. They ultimately harm themselves and certainly harm many others. Risk may also be unintentionally created by actors who single-mindedly pursue a project in ignorance of its potential for causing harm. We saw that mega-projects, such as high dams and mining operations, typically construct risk because they involve large-scale modification of the landscape, heavy engineering, and often cause displacement of resident populations.

Arriving at our present Chapter 10, we saw that vulnerability to disaster risk can be tackled through multilevel, connected governance that begins with the local knowledge and skill of primary residents and risk bearers, who must control the manner and pace of work to construct safety, abundance, and meaning. This work requires movement beyond allegedly universal understandings of disaster and the very concept of disaster risk without losing sight of it. In this way, we will be able to reconsider everyday risks and then to navigate beyond risk by joining with others to create opportunities for living well.

Limitations on Understanding Disaster Risk

Chapter 1 suggested a fundamental limitation related to the philosophical problem of "other minds". We cannot know deeply and with certainty how other people experience risk or what fear or anxiety feel like to them. We have no direct access to others' emotions. We must rely on language that is imprecise and culture bound. Otherwise, we can fall back on psychological, sociological, and anthropological theories that remove us even further from the experience of others.

Chapter 2 introduced another limitation. The "progression of vulnerability" (or "crunch") framework was asserted to be the basis of a solution to the problems posed in Chapter 1. Apart from the imprecision of metaphor, diffusion of the "crunch" (PAR) model has been patchy. Even using the most sophisticated search engines, it is impossible to follow PAR's progress as far as the desks of local government officials and other actors at the scale of primary localities. If action and not just understanding is the goal, then from internet searches only, PAR fails.

Chapter 3 emphasizes the relevance and importance of people's organic abilities/capacities – that is, no one is helpless in facing hazards. People are knowledgeable, resourceful, and creative. Their knowledge, resources, and creativity are often grounded and embedded within cultural practices, norms, and values, which, at the same time, may discriminate against other people and lead the latter to be at greater risk. There is, therefore, a very fine line to walk between maximizing local abilities/capacities and addressing root causes of vulnerability.

Chapters 4 through 8 dealt with specific hazards embedded in fields of human action that often benefited from the same biogeophysical processes that might harm us. We saw that in all cases, be it the nature and behavior of microbes, cyclones, volcanoes, or asteroids, human knowledge is incomplete. We know so little about the forces that hold matter together and give it mass at the tiniest scale and about the most distant galaxies, whose light reaches us billions of years after the Big Bang. We still explore what exactly dark matter and dark energy are; although the former seems to account for most of the mass of the universe and the latter seems to drive its expansion.

Chapter 9 reminded us that "hell is other people", with apologies to Jean-Paul Sartre, as this quotation here refers to both Sartre's original intention and its standard misinterpretation. Intentional and unintentional construction of risk imperils all the projects we aspire to carry forward and our daily lives. As though living on a tectonically active planet with a restless atmosphere and non-human life capable of killing us is not enough, other humans make matters worse. What we think we know about hazard, vulnerability, risk, and disaster can easily be overturned by what a corrupt official does or allows to be done or when the advice of a well-meaning "expert" turns out to be wrong.

Actions Guided by Understanding, Giving Rise to More Questions

This chapter began with a critique of an earlier seven-part call to action. Having updated and modified those objectives, we ended up with signposts that could lead to increased safety and to sustainable livelihoods and more meaningful life. Using the jargon inspired by two editions of the book, *At Risk*, this pathway could lead to the progression of safety (POS) and the progression of life and livelihood (PLL). Our use of the word "could" signals contingency. The manner of pursuing POS and PLL is critical. Few people entirely alone can construct safety, create a sustainable livelihood, or live a meaningful life. Other people are usually required, and they must be involved in choosing the manner and the pace of the changes, as well as the investments and risks undertaken in order to travel this pathway.

The alternative is to drive the construction of safety from the top-down. The top-down approach was soundly rejected by a large number of disaster and development researchers and advocates in 2019 when they helped write and then signed a Manifesto and Accord. The signatories thought in detail about how to carry out disaster studies in a way that is respectful of local people (including ourselves) and build upon everyone's skills, knowledges, and wisdoms. The consensus reached by the signatories can be summarized as follows:

1. Disaster studies should model respect for and trust in local and grounded researchers, as well as their knowledge and abilities, no matter where they come from.
2. Whenever and wherever disasters happen, local and grounded researchers should lead investigations as principal investigators of any research project that deals

with risks and disasters. They should also lead academic and non-academic publications, both oral and written.

3. Outside researchers should come and support locally-driven initiatives only when needed. When such collaboration is warranted, local and grounded researchers and/or local people must retain power in leadership and decision-making.
4. Local and grounded research epistemologies and Indigenous constructs of disasters should be central to disaster studies to better reflect diverse local realities. Researchers should thus value local ontologies and epistemologies, whenever appropriate, to move beyond Enlightenment-based sources, concepts, methodologies, and languages that dominate the field. This does not mean abandoning science in relativism but the respectful acknowledgement of hybridity.
5. Disaster research should address the root causes of vulnerability and recognize the capacities, capabilities, and abilities of local and grounded people. Our research should, therefore, be geared toward reducing the risk of disaster, rather than toward building academic reputation.
6. The dissemination of research outputs and outcomes must occur in a way that demonstrates collaboration, local leadership, and appreciation for local and grounded knowledge and ways of collecting and presenting knowledge. We should share and present knowledge in languages, modes, and mediums accessible to people who can/want to make use of this knowledge.

This way of conducting research can be generalized to apply to the collaborative work of constructing safety and meaning (CSM) and to the pursuit of sustainable livelihood and meaningful living (PLL). Over past millennia, authors have discussed PLL but in more familiar language. They produced narratives and visions of "good work" and "the good life". And, as always, it can and should all be critiqued, challenged, refined, and improved. The Manifesto and Accord are a process.

Good Work

Michael Cholbi (2022) provides an overview of philosophical reflection on work:

> Traditional Confucian thought . . . embraces hard work, perseverance, the maintenance of professional relations, and identification with organizational values. The ancient Mediterranean tradition, exemplified by Plato and Aristotle, admired craft and knowledge-driven productive activity while also espousing the necessity of leisure and freedom for a virtuous life. The Christian tradition contains several different views of work, including that work is toil for human

> sin, that work should be a calling or vocation by which one glorifies God or carries out God's will, and that work is an arena in which to manifest one's status as elect in the eyes of God (the "Protestant work ethic"). The onset of the Industrial Revolution and the adverse working conditions of industrial labor sparked renewed philosophical interest in work, most prominently in Marxist critiques of work and labor that predict the alienation of workers under modern capitalism and the emergence of a classless society in which work is minimized or equitably distributed.

A less philosophical and more action-oriented view of "good work" comes from one of the founders of an institute for building the collaborative capacity in individuals, organizations, and networks seeking social justice and racial equity (Parker, 2012). Parker writes:

> Good work connects your values and your actions. It satisfies a craving to express your purpose, and to do that in ways that align intention and impact. It's about affecting the hearts and minds of individual people, and also about changing outcomes for groups of people. It's about connecting the dots among people and groups, strengthening relationships and networks.

In short, good work means making fundamental changes in how power is constructed and what is considered normative. It is about reordering how business gets done or how institutions function. It's about redefining the "other" so that no one is outside our circle of concern.

Good Life

At a meeting of cultural and political leaders from Somali clans convened when it seemed that peace was imminent, one delegate offered thoughts about people living abundantly and well. Later, these thoughts formed the core of a book chapter (Wisner, 1994) whose subtitle was "Living with the Somali land, and living well". That aspiration and potential to live with each other and with the land harmoniously can be generalized to all humans on Earth. As the principles of the Manifesto and Accord specify, there is no single universal way of achieving that. To construct safety and PLL will take many pathways determined by local people cooperating with each other. In fact, in their daily lives, people are continuously attempting unselfconsciously to construct safety and PLL.

A common Western view is that the good life is the absence of pain and satisfaction of desires (Arthur Schopenhauer) or pleasure and freedom from pain in John Stuart Mill's view. By contrast, Maimonides, a medieval Jewish philosopher, argued that perfection of the intellect was the point of living. Interestingly enough, two modern women believe that the point of life is the living of it (Eleanor Roosevelt) or, in the words of Virginia Woolf, to enjoy "little daily miracles, illuminations, matches struck unexpectedly in the dark" (1930, p. 239). The Dalai Lama's view is more social: the purpose of life is to be happy, and happiness requires kindness and open-heartedness toward others. A great diversity emerges across cultures and through time. Each of us has to find our own path in the search for safety, livelihood, and meaningful living, hoping that it does not interfere too much with others' searches and discoveries.

Conclusions

Reversing the progression of vulnerability is not easy because PAR is so far only partially understood. There are limitations to our understanding of disaster risk, so it must remain unclear – pending future clarity – just what it is that has to be "reversed" in order to produce a progression of safety (POS). In any case, merely reversing PAR is not enough to construct safety. Safety requires a context that involves sustainable livelihood and a state of mind that is free of anxiety and capable of living life joyfully with other people and non-human living beings. The specific form of good work and the good life that results will carry the imprint of the diverse cultures. The pace and manner in which the progression of safety and meaningful life are pursued must involve those affected. We must all respect the skills, knowledges, and wisdoms of each other, applying local and grounded methods of work. Mutual trust and respect are essential.

Note

1 Wisner (1988) discusses a basic human needs approach to development (BNA) and notes that individuals alone cannot satisfy them, nor can they be satisfied by some external authority, but require what he terms "transformational" participation by a group of people who thereby define needs and the ways they are to be satisfied. This argument rescue's the BNA from the solipsism of an individual trapped in Hobbes' (1651) "war of all against all" and the authoritarianism of experts seeking to define the needs of others (Foucault, 1976).

References Cited

Bello, W. (2004). *Deglobalization.* Zed.

Blaikie, P., Cannon, T., Davis, I., & Wisner, B. (1994). *At risk: Natural hazards, people's vulnerability and disasters* (1st ed.). Routledge.

Butler, J. (2009). *Frames of war: When is life grievable*? Verso.

Butler, J. (2022). *What world is this? A pandemic phenomenology*. Colombia University Press.

Cholbi, M. (2022). Philosophical approaches to work and labor. In E. Zalta & U. Nodelman (Eds.), *The Stanford encyclopedia of philosophy* (Winter 2022 ed.). Stanford University.

Corlett, R. (2020). Safeguarding our future by protecting biodiversity. *Plant Diversity, 42*(4), 221–228.

Foucault, M. (1976). *La volonté de savoir*. Gallimard.

Gaillard, JC., & Mercer, J. (2013). From knowledge to action: Bridging gaps in disaster risk reduction. *Progress in Human Geography*, 37(1), 93–114.

Harris, B., & Ingizza, C. (2023, April 10). Brazil workers' movement steps up land invasions under Lula government. *Financial Times*. https://www.ft.com/content/2852dec2-ed95-4c53-b68b-0085913ddf04

Health and Safety Commission. (1993). *Third report: Organizing for safety*. Health and Safety Commission.

Hobbes, T. (2010). *Leviathan or the matter, forme and power of a Commonwealth Ecclesiasticall and Civil*. Yale University Press. (Original work published 1651)

Latour, B. (2017). *Facing Gaia: Eight lectures on the new climate regime*. Polity.

Latour, B. (2018). *Down to earth: Politics in the new climate regime*. Polity.

Latour, B. (2021). *After lockdown: A metamorphosis*. Polity.

Lawler, O., Allan, H., Baxter, P., Castagnino, R., Corella, M., Dann, L., Hungerford, J., Karmacharya, D., Lloyd, T. J., López-Jara, M. J., Massie, G. N., Novera, J., Rogers, A. M., & Kark, S. (2021). The COVID-19 pandemic is intricately linked to biodiversity loss and ecosystem health. *The Lancet, 5*(11), e840–e850. https://doi.org/10.1016/S2542-5196(21)00258-8

Marquardt, J., & Lederer, M. (2022). Politicizing climate change in times of populism: An introduction. *Environmental Politics, 31*(5), 735–754. https://doi.org/10.1080/09644016.2022.2083478

Mia, I. (2022). *The armed conflict survey 2022*. International Institute for Strategic Studies.

Molière, J.-B. P. (1670). *Le Bourgeois Gentilhomme [The middle class gentleman]*. Presented at Chambord, for the Entertainment of the King.

Parker, C. (2012, October 12). *What is good work*? Interaction Institute for Social Change. https://interactioninstitute.org/what-is-good-work/

Peter, M., Diekötter, T., Höffler, T., & Kremer, K. (2021). Biodiversity citizen science: Outcomes for the participating citizens. *People and Nature, 3*(2), 294–311. https://doi.org/10.1002/pan3.10193

Singer, P. (2023). *Animal liberation now*. Harper Perennial.

Teilhard de Chardin, P. (2008). *The phenomenon of man*. Harper Perennial.

United Nations Office for Disaster Risk Reduction. (2022). *Our world at risk. Global assessment of disaster risk reduction 2022*. United Nations Office for Disaster Risk Reduction.

Watson, J. E., Venter, O., Lee, J., Jones, K. R., Robinson, J. G., Possingham, H. P., & Allan, J. R. (2018). Protect the last of the wild. *Nature, 563*(7729), 27–30. https://doi.org/10.1038/d41586-018-07183-6

Wisner, B. (1988). *Power and need in Africa: Basic human needs and development policies.* Earthscan.

Wisner, B. (1994). Jilaal, gu, hagaa, and der: Living with the Somali land, and living well. In A. Samatar (Ed.), *The Somali challenge* (pp. 25–64). Lynne Rienner.

Wisner, B., Blaikie, P., Cannon, T., & Davis, I. (2004). *At risk: Natural hazards, people's vulnerability and disasters* (2nd ed.). Routledge.

Wisner, B., Gaillard, JC., & Kelman, I. (2012). Framing disaster: Theories and stories seeking to understand hazards, vulnerability and risk. In B. Wisner, JC. Gaillard, & I. Kelman (Eds.), *The Routledge handbook of hazards and disaster risk reduction* (pp. 18–33). Routledge.

Woolf, V. (1930). *To the lighthouse.* Hogarth Press.

World Inequality Report 2022. (2022). *World inequality lab.* https://wir2022.wid.world/www-site/uploads/2023/03/D_FINAL_WIL_RIM_RAPPORT_2303.pdf

Wynberg, R. (2023). Biopiracy: Crying wolf or a lever for equity and conservation? *Research Policy, 52*(2), 104674. https://doi.org/10.1016/j.respol.2022.104674

Afterword: Head and Heart – Compassion and Anger in Disaster Studies and Action

As we come to the end of this book, we wonder: Will it make any difference in a world where there is so much violent conflict and so many corrupt individuals, governments, and institutions? We feel compassion for those caught up in war and silent/quiet/slow violences, anger in the face of corruption, laziness, and malfeasance on the part of office-holders and corporate and nonprofit leaders whose responsibility is people's safety. We also feel anger when witnessing how people with abilities and local knowledge who could act to protect themselves and others are blocked or hampered by self-interested, powerful individuals or oppressive institutions or simply because of racist, sexist, or elitist disregard for what ordinary people know and think. Weariness comes because of déjà vu. We have seen corruption, incompetence, and dependency for more than 50 years since the current era of "disaster management" began with the creation of the Office of the United Nations Disaster Relief Coordinator.

Scoundrels in government and other positions of power may claim that they had no warning, were surprised, and were overwhelmed by the size of the hazard impact. This is just one of many possible excuses for inaction. Heads of state, heads of government, and their officials and apologists may blame the overwhelming magnitude of earthquakes and storms. Yet countries with known seismic faults or flood hazards should have adopted appropriate building standards and zoning regulations as well as systems for monitoring and enforcement. Those with power must bear responsibility. Action and inaction must be seen clearly to have consequences. We reject such excuses. Climate change is well understood, and governments should, by now, have incorporated the results of climate change science, such as increased storm intensity, into their regulations and plans.

Of course, we recognize that a reference to the extreme magnitude of hazards may not always be an attempt to excuse a lapse of responsibility. It can also function to mobilize humanitarian support. Even in such cases, unmentioned causes and vulnerabilities turn a hazard into a disaster. For example, the landslide disaster on 13 March 2023 in Manaus, Brazil, could be explained by the urban sprawl triggered by the creation of a free-trade zone in the Amazon rainforest, a maldevelopment project led by the military dictatorship in the 1970s (Marchezini & Wisner, 2017). In 1985, Manaus had 523 hectares of informal settlements in highly hazard-prone areas, while in 2021, those increased to 1,841 hectares (Lacerda, 2023). Many people live in stilt houses or on steep slopes, including about 40,000 people who migrated from Venezuela to Manaus due to political and economic crises (Portal G1, 2023). Of eight people lost in this landslide disaster, four were Venezuelan.

Tools and frameworks are needed to peel back excuses and to investigate the root causes of vulnerability in order to reveal how disaster risk is produced and reproduced. The resulting revelatory and documented understanding would serve as a means to clarify research questions and directions. It would be a vehicle for teaching, and it could galvanize effective policy and action. Such tools could be used by watchdog organizations and lobbyists for leadership accountability.

This afterword interlaces compassion, weariness, and anger with an attempt to frame and understand who is responsible for the processes of disaster risk construction. We have seen for decades the same questions asked and the same proposals repackaged in international initiatives every five to ten years. There has been some progress. There is much more awareness of the social factors that influence vulnerabilities. Programs, practices, and activities on the ground have become more inclusive across demographic and identity variables. Yet who remains marginalized, excluded, or underrepresented? Some trends suggest decreased disaster mortality, although the data are generally inadequate for such calculations, and these figures do not factor in disease, most recently the COVID-19 pandemic that killed millions alongside the huge numbers dying each year from other pathogens and parasites, such as malaria, which alone leads to over half a million deaths annually (WHO, 2023).

These are the reasons for our anger – anger because we know exactly what to do in order to avoid disasters and yet those with the power and position to act are demonstrably talented at failing to apply this knowledge. We feel compassion for fellow human beings being harmed through no fault of their own – accepting that, any day, it could be any of us or our loved ones. This is the condition of daily life. There continues to be avoidable death, injury, and disruption – that is, calamity – simply because those with power do not offer the knowledge, money, opportunity, and choices to stop and reverse these processes of disaster risk construction.

Our compassion and anger grew rapidly during and after October 2023. First, Hamas brutally slaughtered almost 1,200 people, mainly civilians, in Israel and took over 230 hostages, again mainly civilians. Then the Israeli army bombarded and invaded Gaza, killing tens of thousands of civilians. In the meantime, the horrific casualty toll of Russia's full-scale invasion of Ukraine continues to rise, while Afghanistan, Yemen, the Democratic Republic of Congo, and Nigeria are among the many other countries where civilians are routinely targeted during conflict.

Yet there is always positive news to share, to inspire us and to show us ways forward. After the 6 February 2023 earthquakes along the Türkiye-Syria border that took over 60,000 lives, collapsed structures were frequently ringed by intact buildings. Clearly, we can keep buildings upright during earthquakes, and the worst-hit areas of Türkiye had done so for many. Bangladesh, too, awaits the horror of a mass-casualty earthquake yet has reduced cyclone deaths from tens of thousands per storm to dozens. Toronto, Ontario, and Boulder, Colorado, have moved dozens of homes out of floodplains, saving lives on numerous occasions when major flooding made the ravines dangerous but did not lead to a disaster since no one was in the way of the floodwater. Disasters in Toronto and Boulder are far from precluded, but flood vulnerabilities have been substantially reduced. The speed of COVID-19 vaccine development – notwithstanding major distribution and management problems, alongside over-optimistic claims about vaccine effectiveness while downplaying some side effects – is a testament to what humanity can do when we come together and offer resources for saving lives rather than taking lives. There are positives glimmering among the ever-present negatives.

Impetus and motivation for positive actions can be provided through legal and administrative requirements, financial incentives and disincentives, and social and political pressures. These measures can and should be complemented by attempts to change the consciousness of those with power and resources by emotional appeal and specific stories of successes and failures. In addition to explaining again and again, step by step, that disaster risk emerges from a combination of vulnerability, hazard, and lack of preventive and mitigative policies to address them, we could engage the powerful with an invitation to join with us in simple, straightforward terms: "We can save lives and improve them. Will you join us in doing so?"

References Cited

Lacerda, P. (2023, March 18). *Manaus sofre com expansão urbana em assentamentos precários.* Agência Brasil. https://agenciabrasil.ebc.com.br/geral/noticia/2023-03/manaus-sofre-com-expansao-urbana-em-assentamentos-precarios

Marchezini, V., & Wisner, B. (2017). Challenges for vulnerability reduction in Brazil: Insights from the PAR framework. In V. Marchezini, B. Wisner, L. R. Londe, & S. M. Saito (Eds.), *Reduction of vulnerability to disasters: From knowledge to action* (pp. 57–96). Editora Rima. http://www2.cemaden.gov.br/publicacoes-cemaden/

Portal G1. (2023, March 13). Quatro venezuelanos da mesma família estão entre os mortos em deslizamento de terra em Manaus. *Portal G1*. https://g1.globo.com/am/amazonas/noticia/2023/03/13/quatro-venezuelanos-da-mesma-familia-estao-entre-os-mortos-em-deslizamento-de-terra-em-manaus.ghtml

WHO (World Health Organisation). (2023). The 2023 WHO world malaria report. *The Lancet Microbe*. https://www.thelancet.com/journals/lanmic/article/PIIS2666-5247(24)00016-8/fulltext#:~:text=An%20estimated%20608%20000%20deaths,100%20000%20population%20at%20risk

Index

Note: Page numbers in *italics* indicate a figure and page numbers in **bold** indicate a table on the corresponding page. Page numbers followed by "n" with numbers refer to notes.